Flaws and Fallacies in Statistical Thinking

Stephen K. Campbell

University of Denver

PRENTICE-HALL, INC.

Englewood Cliffs, New Jersey

Library of Congress Cataloging in Publication Data

CAMPBELL, STEPHEN KENT.
  Flaws and fallacies in statistical thinking.

  Includes bibliographical references.
  1. Statistics. I. Title.
HA29.C253   001.4'22   73–5655
ISBN 0–13–322214–4

10 9 8 7 6 5 4 3 2

PRENTICE-HALL INTERNATIONAL, INC., *London*
PRENTICE-HALL OF AUSTRALIA, PTY. LTD., *Sydney*
PRENTICE-HALL OF CANADA, LTD., *Toronto*
PRENTICE-HALL OF INDIA PRIVATE LIMITED, *New Delhi*
PRENTICE-HALL OF JAPAN, INC., *Tokyo*

To Judy

Contents

Preface

When a person loves something he hates to see it abused. Such is my feeling toward statistics. Hardly a day goes by that I fail to see the subject I love put to use in faulty and misleading ways in newspapers, magazines, books, speeches, and, especially, advertising.

Curiously, little of the statistical literature is aimed at helping the nonstatistician recognize such abuses. Statisticians, understandably, tend to write for other statisticians. Still, one of life's rare truisms is that most people in the world are not statisticians and don't aspire to become statisticians. Nevertheless, it is an ususual person who isn't frequently confronted with the need to evaluate, if only very informally, statistical information when making decisions about a variety of subjects ranging from which political candidate to vote for to which brand of detergent to buy. Professional statisticians don't need help along such lines; keen critical judgment about statistical information is, for most, a natural by-product of their professional training. But who is to help the statistical layman if it is not the statistical professional?

This book was written with two purposes in mind—the first, less important, purpose being that of getting something off my chest. For many years I have been distressed by the frequency with which (1) relatively simple statistical tools such as percents, graphs, and averages are misused and (2) faulty conclusions are drawn from, perhaps, flawless data in our news media simply because the purveyors of the information don't know any better. Moreover, I have been annoyed—indeed, made damned mad—by the frequency with which bogus statistical evidence is used intentionally by some unconscionable people to sell their products or pet ideas to others. Writing this book has been good therapy for me.

The second, more important, purpose of this book is loosely related to the first. I have long felt that the university student who is likely to take only one or two courses in statistics and the ordinary citizen who maybe lacks any classroom exposure whatever to the subject could benefit from a nontechnical book written with a view to helping him increase his ability to judge the quality of statistical evidence and, in turn, to make better-informed decisions about many facets of everyday life. This book, therefore, has been written both as a supplemental reading text for the student taking his first course in statistics and as a self-help guide for the nonstudent who feels the need to evaluate statistical evidence more judiciously than he is presently capable.

The sequence of topics covered is in rough accord with that of many beginning textbooks in statistics. The terminology used and the manner in which the subjects are treated are based on the assumption that the reader has had little or no prior exposure to the subject of statistics and has studied precious little formal mathematics.

Many people have helped me with the preparation of this book—more than I can possibly thank individually. I must, however, single out three statistics professors for special thanks. Professors Richard B. Ellis of Northern Essex Community College and Richard E. Lund of Montana State University both reviewed early drafts of the first five chapters and offered many incisive criticisms and imaginative suggestions for improving upon my proposed project. My colleague and friend, Professor Paul R. Merry of the University of Denver, went over the final manuscript with a thoroughness that is so characteristic of everything he does and offered numerous constructive suggestions.

I would also like to thank my statistics students at the University of Denver who gathered hundreds of examples of statistical fallacies from which many of those appearing in the following pages were selected. Also deserving of thanks are several authors, editors, and advertisers who granted me permission to quote from copyrighted material even though they must have suspected that I intended to be more critical than complimentary.

Last, but certainly not least, I wish to thank my wife Judy who not only relinquished without complaint many hours of time with me which were rightfully hers but who also did more than her fair share to create an atmosphere within which the work could proceed with a bare minimum of discord or distraction.

Needless to say, any blame for errors, omissions, or bad manners should be directed at me.

STEPHEN K. CAMPBELL

1

Dangers of Statistical Ignorance

Statistical thinking will one day be as necessary for efficient citizenship as the ability to read and write.

—H. G. WELLS

This book is unusual. Textbooks show you facts and the right methods. This book shows you fallacies and the wrong methods. It will serve as a companion volume to any textbook on statistics. It will also serve as a self-help guide for the layman who desires to distinguish between valid and faulty statistical reasoning.

Furthermore, it deals with a very important subject because statistics influence our daily lives in a great many ways. Through statistics, twentieth century man measures economic activity; records social progress; elects Presidents and keeps abreast of their current popularity (or, more often, unpopularity); measures intelligence, interests, and aptitudes; compares his sexual habits with various norms; determines which television shows will survive and which will not; compares the profit potential of several alternative business strategies; decides whether to invest in bonds or stocks and, if the latter, whether now is a good time to get into the market; keeps track of batting averages; assesses the likelihood of rain tomorrow; and, in general, keeps informed about what is going on in the world with the aid of statistical data gathered, presented, and interpreted by others. Even if you and I have nothing to do with the actual calculations implied by the items in this list (or the many other items that might easily have been included), as socially conscious citizens we should be able to interpret the results of such calculations with some sophistication, for these are the figures that serve as a basis for so many vital newspaper and magazine articles, books, and speeches.

Although something of an exaggeration, the quote from H. G. Wells that introduced this chapter is basically sound. I would amend it only in one important respect and have it read "Statistical [straight] thinking will one day be as necessary for efficient citizenship as the ability to read and write." I prefer "statistical straight thinking" to "statistical thinking" because it seems unlikely that fuzzy or erroneous thinking could contribute much to efficient citizenship—or efficient anything else, for that matter. Unfortunately, there is enough fuzzy and erroneous statistical thinking around these days to justify my focusing on it in this book.

This book deals with erroneous and sometimes deliberately misleading statistical arguments. It deals with fallacious statistical thinking—how to avoid doing it yourself and how to recognize when others do it. This point requires elaboration. First, however, let us touch on some essential background topics, not the least important of which is what is meant by *statistics*.

The Two Meanings of "Statistics"

What is or are statistics? The word has two widely used meanings. The most generally familiar—and for many people the least interesting—can probably be introduced most painlessly by the following excerpt from O. Henry's *Handbook of Hymen*:

> "Let us sit on this log at the roadside," says I, "and forget the inhumanity and ribaldry of the poets. It is in the columns of ascertained facts and legalized measures that beauty is to be found. In this very log we sit upon, Mrs. Sampson," says I, "is statistics more wonderful than any poem. The rings show it is sixty years old. At the depth of two thousand feet it would become coal in three thousand years. The deepest coal mine in the world is at Killingworth near New Castle. A box four feet long, three feet wide, and two feet eight inches deep will hold one ton of coal. If an artery is cut compress it above the wound. A man's leg contains thirty bones. The Tower of London was burned in 1841."
>
> "Go on, Mr. Pratt," says Mrs. Sampson, "Them ideas is so original and soothing. I think statistics are just as lovely as they can be."

Although not all of Mr. Pratt's original and soothing ideas are really statistics, enough of them are to convey the idea that a statistic is a fact. More precisely, it is a fact expressed as a number and can be a measurement (an attractive girl's 36-24-36 "statistics," for example), a count (the number of girls having blond hair in a beauty contest), or a rank (a specific girl's standing in a beauty contest, the winner being assigned a rank of "1"). A statistic in this first sense can even be a summary measure such as a total, an average, or a percentage of several such measurements, counts, or ranks.

In addition to referring to numerical facts, the term "statistics" also applies to the broad discipline of statistical manipulation in much the same way that "accounting" applies to the entering and balancing of accounts. "Statistics" in this broader sense is a set of methods for obtaining, organizing, summarizing, presenting, and analyzing numerical facts. Usually these numerical facts represent partial rather than complete knowledge about a situation, as is the case when a sample is used in lieu of a complete census. Generally speaking, numerical facts are subjected to formal statistical analysis in order to help someone make wise decisions in the face of uncertainty or to help researchers arrive at scientifically-sound generalizations or principles.

The word "statistics" will be used in both senses throughout this book. The context within which the term is used should make the intended meaning clear.

The Statistical Fallacy

No one knows just when the first statistical lie was foisted upon a trusting listener. For that matter, no one knows for certain when or where statistics first appeared. We do know that the earliest written records contain numbers, a fact suggesting that the ability to count goes way back. The Bible tells us that statistics in the purely descriptive sense were used to provide information about taxes, wars, agriculture, and even athletic events. Nevertheless, there probably was a time when counting, and therefore statistics, was unknown; a time when a shepherd, for example, did not describe his flock as consisting of twenty, fifty, or one hundred sheep but instead kept track of his woolly charges by assigning each a name. If two sheep turned up missing, the shepherd searched not for two anonymous animals but for, say, Peter and Paul.

Although the first uses of statistics are lost in antiquity, I would wager that misuses of statistics—intentional and unintentional—first appeared at about the same time as valid statistics. We all know people whose honesty we have good reason to doubt as well as people who are chronically careless or conspicuously stupid. Certainly there must have been counterparts of such people from the very beginning just as there must have been counterparts of you and me—honest, meticulous, and highly intelligent. Not much imagination is required to envision our shepherd, recounting the harrowing challenges he faced while retrieving his two wayward sheep, Peter and Paul, and, rather than spoiling a good story by undertelling it, claiming that Mary and Ruth had also gone astray (had gotten lost, that is). The advent of counting and statistics certainly didn't create the all-too-human tendencies to lie, exaggerate, or make honest mistakes, but it did introduce a whole new, very colorful means of giving vent to such tendencies.

Today, statistical fallacies abound in our newspapers, magazines, advertisements, and conversations. I am not suggesting that all statistical evidence is faulty. Indeed, the proliferation of statistical fallacies in recent decades has been, to a considerable extent, the natural result of burgeoning statistical data and formal techniques for analyzing such data. But even so, the mere existence of statistical fallacies imposes a responsibility upon the citizen who would call himself well informed to learn to distinguish between erroneous and valid statistics or statistical arguments.

How dangerous are statistical fallacies? No general answer is possible. Some statistical fallacies are undoubtedly perfectly harmless even when widely believed. But some are much more potentially dangerous than you might suppose. Let us consider a few examples.

If an award is ever granted the fabricator of the world's phoniest but possibly least harmful statistic of a descriptive nature, I suggest that the honor be bestowed posthumously upon a German named Weirus. Weirus, who served as physician to the Duke of Cleaves during the latter part of the sixteenth century, a time when most of Europe was gripped by the fear of demons and witches, had some definite opinions about the number of demons in existence. Whereas most of his contemporaries lazily assumed that demons were too plentiful for their numbers to be determined, Weirus, using methods beyond anyone else's comprehension, calculated that exactly 7,405,926 demons inhabited the earth; these, he claimed, were divided into seventy-two battalions, each under a prince or captain.[1]

How serious Weirus was when he revealed his remarkable findings is anybody's guess. According to Sir Walter Scott, "Weirus was one of the first

---

[1] Joseph Jastrow, *Error and Eccentricity in Human Belief* (New York: Dover Publications, Inc., 1962), p. 86.

who attacked the vulgar belief and boldly assailed, both by serious argument and by ridicule, the vulgar credulity on the subject of wizards and witches."[2] Quite possibly the good doctor had his tongue planted firmly in his cheek when he revealed his spurious figures. If so, many historians have not been in on the joke for they have written of the incident as if they believed Weirus to be dead serious. Moreover, you can bet that many people hearing of the doctor's calculations accepted the bogus results without question, thinking that 7,405,926 sounded "about right." And why not? To them demons were a reality and Weirus was a learned man. In the final analysis, however, it is hard to imagine how Weirus' figures could have done anyone harm, except perhaps Weirus himself if he really was poking fun at the prevailing beliefs of his day.

Here is a more modern-day example. A 1968 advertisement for Volvo automobiles is a treasure trove of statistical fallacies, but that unfortunate fact didn't keep it from appearing in many of the country's top magazines. The advertisement states that, according to statistics, the average American drives 50 years in his lifetime and the average car is traded in on a new one every three years and three months. The logical conclusion, we are informed, is that if one drives an average number of years in average cars, he will own

[2] *Letters On Demonology and Witchcraft* (London: W. Tegg, n.d.), p. 186.

15.1 cars in his lifetime. But not so if he owns a Volvo. The lucky Volvo owner, the ad asserts, can get by with only 4.5 Volvos because in Sweden Volvos last an average of 11 years.

The advertisement goes on to say, "We don't *guarantee* they'll last that long here where being a car is relatively easy. But we do know that over 95% of all Volvos registered here in the last 11 years are still on the road."

The Volvo may indeed be a durable, well-constructed automobile—I have no convictions one way or the other—but, despite the collection of presumably authentic figures, this advertisement conveys no genuine information about the Volvo's durability relative to that of other cars.

Aside from the question of whether the number of years a car is driven is the most meaningful possible measure of durability (why not number of miles driven instead?), two clear-cut fallacies stand out. First, an improper comparison is made. The almost adjacent statements "The average car is traded in on a new one every three years and three months" and "Volvos last 11 years . . ." is an apples and oranges comparison if there ever was one. Obviously, the number of years an automobile can be driven before it becomes incapable of providing transportation and the number of years an automobile is driven before the owner tires of it and voluntarily trades the still usable vehicle in on a newer model are two very different matters. No meaningful comparison whatever can be made between the two figures.

Second, the last part of this ad—the part that states " . . . over 95% of all Volvos registered here in the last 11 years are still on the road"—is no more informative than the first part and is just as potentially misleading. Suppose, for example, that all Volvo sales made during the 11 year period referred to (apparently 1957 through 1967 inclusive) had been made during the most recent year and none whatever during the preceding ten years. In such a case, the figure of 95 percent would be indicative of poor rather than outstanding durability because it would mean that five percent of the Volvos sold had to be scrapped during their first year of use. Actual sales admittedly were not bunched so dramatically, but, according to a 1968 issue of *Ward's Automotive Reports*, approximately 45 to 50 percent of the Volvo sales in this country were made during the most recent four years of the 11-year period in question. The 95 percent figure, therefore, says little or nothing about the Volvo's durability.

What can we conclude about this advertisement other than it is much more misleading than informative? Is it dangerous? That is quite possibly a different matter altogether. If the Volvo is in fact an unusually durable car, then the advertisement presumably did no real harm. (Of course, one can hardly help wondering why, if the car really is all that durable, the company has to resort to half-truths to sell it.) But if the Volvo really is no more durable than other makes, and certainly if it is less durable, then many automobile buyers might have been led astray by the ad, and, according to any criteria I can think of, the statistical fallacies that helped to sell the car would have to be viewed as dangerous.

Volvo says:

"95% OF THE VOLVOS REGISTERED IN THE U.S. DURING THE PAST 11 YEARS ARE STILL ON THE ROAD".

Now let us consider two heavyweights. These are examples of statistical fallacies whose propensities for causing trouble are incalculably great. The first example comes from a *Playboy* Magazine interview with the late George Lincoln Rockwell, commander of the American Nazi Party:[3]

> Rockwell: A psychologist named G.O. Ferguson made a definitive study of the connection between the amount of white blood and intelligence in niggers. He tested all the nigger school children in Virginia and proved that the pure-black niggers did only about 70 percent as well as the white children. Niggers with one white grandparent did about 75 percent as well as the white children. Niggers with two white grandparents did still better and niggers with *three* white grandparents did almost as well as the white kids. Since all these nigger children shared exactly the same environment as niggers, it's impossible to claim that environment produced these tremendous changes in performance.
>
> Playboy: In his book, *A Profile of the Negro American*, the world-famed sociologist, T. F. Pettigrew states flatly that the degree of white ancestry does not relate in any way to Negro I. Q. scores. According to Pettigrew, the brightest Negro yet reported—with a tested I. Q. of 200—had no traceable Caucasian heritage whatever.
>
> Rockwell: The fact that you can show me one very black individual who is superior to me doesn't convince me that the average nigger is superior. The startling fact I see is that the lighter they are, the smarter they are, and the blacker they are, the dumber they are.

Rockwell's faith in the Ferguson study might be rather touching if the moral stakes weren't so high and if the study enjoyed any scientific repute,

---

[3] "Playboy Interview: George Lincoln Rockwell," *Playboy*, April, 1966, pp. 71ff.

a consideration that Rockwell didn't bother to worry about. But as the magazine's editors point out:

> Ferguson's study, conducted in 1916, we later learned, has since been discredited by every major authority on genetics and anthropology; they call it a pseudoscientific rationale for racism, based on inadequate and unrepresentative sampling, predicated on erroneous assumptions, and statistically loaded to prove its point.

Credit for the second heavyweight statistical fallacy goes to Joseph Stalin and concerns statements he made about the success of his first Five Year Plan. The story is related most colorfully in Eugene Lyons' *Workers' Paradise Lost*:[4]

> No other economic enterprise in history has been so vastly publicized, so glamorized and misjudged, as Stalin's first Five Year Plan. As originally charted, the Plan covered every department of the nation's life, promising great advances in consumer industries, food production, housing. Meticulously the planning agency, Gosplan, detailed higher living standards. The purchasing power of the Soviet currency would rise by 20 percent, real wages by 66 percent, the cost of living would be lowered by 14 percent.

Lyons continues by describing a speech Stalin himself made only eighteen months prior to the end of the five-year period, a speech in which he came very close to admitting that the Plan had proved a dismal failure. Nevertheless, eighteen months later, in January of 1933, Stalin announced the quantitative fulfillment of 93.7 percent of the entire Plan! What kind of statistical trickery is reflected in this figure? Lyons explains as follows:

> ...The Kremlin simply compared total result with the total planned instead of weighing the actual increase against the *planned* increase. For example, steel output in 1928 was 4.2 million tons. The Plan foresaw an increase to 10.3 million tons. Actual production in the final year was 5.9 million tons—up 1.7 million instead of 6.1 million, or 28 percent of the planned expansion.
>
> The Kremlin, however, said in effect: "We aimed at 10.3 and got 5.9, therefore, out Plan was fulfilled by 57 percent." On this basis, if production had not increased by a single ton, the Plan would have been carried out by over 40 percent—progress while standing still!
>
> When such sleight of hand is revealed, the official claims collapse. New housing, credited with 84 percent fulfillment, in fact increased only 44 percent. ... The actual increase in cement was 37 percent, in brick 28 percent, in

---

[4] (New York: Funk & Wagnall, 1967.) Quotations presented here are from the *Reader's Digest* condensed version of the book (November 1967, pp. 233ff.). The wording but not the essence of the message differs slightly from the original.

automobiles 13 percent. Meanwhile, living costs zoomed, wages declined, hunger spread, consumer goods were tragically short.

Lyons' summary of the situation just described is a ringing testimonial to the potential treachery of a statistical lie when it is told by a strategic political figure at a strategic point in world history. Lyons concludes:

> But amazingly, the Plan has gone down in history as a fabulous success. Indeed, the belief that Communism is a virtual guarantee of rapid economic progress for underdeveloped nations stems primarily from this stubborn delusion which began when Stalin's boasts were accepted across a large part of of the world.

Clearly, faulty statistical reasoning can be relatively harmless or extremely dangerous. Much depends upon what the figures measure, how they are interpreted, and how the conclusions are acted upon.

What Is a Statistical Fallacy?

"The question is," said Humpty Dumpty to Alice, "which is to be master— that's all . . . . When I use a word it means exactly what I choose it to mean —neither more nor less." This, I must confess, is very nearly the same dictatorial attitude I have assumed while culling through many hundreds of statistical arguments to determine the fallacious ones. If an example impressed *me* as a fallacy I labeled it such. Period. In this respect, I have been a little like Lewis Carroll's Humpty Dumpty and the baseball umpire who insisted that a pitch is neither a ball nor a strike until he calls it one or the other.

The task of identifying statistical fallacies is fun but frustrating because the debatable ones and the borderline cases probably outnumber the clear-cut ones. For this reason, I have not attempted to define "statistical fallacy" in a rigorous way and then religiously stick with that definition throughout this book. Maybe a look at the dictionary definitions of *fallacy* will convince you of the impossibility of such as approach:

> 1 (a): Guile, trickery (b): deceptive appearance: deception. 2 (a): A false idea (b): erroneous or fallacious character: erroneousness. 3: An argument failing to satisfy the conditions of valid and correct inference.

Quite a few things, therefore, can be placed under this definitional umbrella. But who is to say what is deceptive and what isn't? In order for something to be deceptive, someone must presumably be deceived. Or is it enough that someone might possibly be deceived? Graphs with broken vertical axes, for example, are often labeled fallacious in articles about misuses of statistics on the grounds that they can give the viewer a false impression of both the level and the rate of change in the data charted. But if such a graph were prepared by a reputable economist for presentation before a group of fellow economists, especially if it were constructed with a view to highlighting important short-term fluctuations, it certainly couldn't be very deceptive.

Similarly, who is to say what is a false idea and what isn't? Sometimes a false idea is very easy to recognize; but when doubt exists, by what infallable criteria is such doubt erased? Consider, for example, this excerpt from a widely read magazine article:

> The production of pornography is a $19-million annual business in my state [California]. Nationwide, the production and sale of pornography is perhaps a $500-million industry. I estimate that more than 50,000 Californians participate in some way in the filth racket.[5]

These assertions can be faulted on the grounds that nowhere in the article is pornography defined. In the absence of either a generally accepted or a Friendly Definition[6] of this colorful word, the specific figures cited are meaningless. Nor are they made more meaningful by their proximity to such vague phrases as ". . . perhaps a $500-million industry" or ". . . participate in some way. . . ." But are the figures false? That, I suppose, depends on what the author is calling pornography—and he is keeping that a carefully guarded secret. I feel that less secrecy would have been desirable and am most willing to attach the label "fallacy" to this example. Someone else, on the other hand, might take the position that "Any darn fool knows what pornography is."

In the final analysis, determining what is and what is not a statistical fallacy will often involve differences of opinion. But perhaps that is all to the good from your standpoint. Since the principal goal of this book is to help you sharpen your critical judgment concerning statistical evidence, you should be prepared to question my position on any or all examples presented in the following chapters.

A Note About the Examples

Not only are statistical fallacies difficult to distinguish from the borderline cases, they are hard to categorize as well. Because the categories often used are not mutually exclusive, a single example can illustrate measurement

---

[5] Max Rafferty, "Crack Down on the Smut Kings!" *Reader's Digest*, November 1968, p. 98.

[6] By *Friendly Definition* I mean a definition selected from among several contending possibilities. The user of the data is in effect asked to accept that specific definition when interpreting the data. In return, the supplier of the data promises to adhere rigorously to that definition. Friendly Definitions are discussed more fully in Chapter 2.

problems, spurious accuracy, faulty comparisons, and maybe several other kinds of fallacies as well.

The difficulties of categorizing statistical fallacies are compounded when the would-be categorizer relies heavily (as I have done) upon examples from real life rather than upon hypothetical examples tailor-made to illustrate a single point in an exaggerated way. Admittedly, I have not eschewed the use of hypothetical examples altogether any more than I have held myself morally aloft from stealing examples from other authors. For the most part, however, the examples presented have been selected from among many hundreds of similar real-life examples collected by myself and several of my former statistics students (although I have admittedly taken the liberty of disguising some for reasons that should be obvious). The examples are drawn from real situations and relate to such diverse subjects as business, economics, psychology, biology, education, sports, entertainment, law, politics, and a great many others. My goal in using this approach has been not only to acquaint you with common misuses of statistics but also to convince you—subliminally if all else fails—that statistics does indeed play a vital role in today's world.

To some extent, the virtues of the examples selected are also their weaknesses. Here are a few caveats:

First, for reasons already given, the categories shouldn't be taken too seriously.

Second, when discussing a particular example, I have usually limited my comments to points bearing directly upon the subject of the chapter or section in which the example appears. Because the same example could perhaps be used to illustrate other kinds of fallacies as well, you may find some of my discussions lacking in thoroughness. Rather than being annoyed by this failing, however, why don't you take it upon yourself to supply the desired thoroughness? It should be good practice.

Third, my approach has necessitated much quoting or paraphrasing out of context. In no case, however, have I intentionally misrepresented another's argument.

In Defense of Being Negative

I promised earlier to elaborate on my statement that this book is about fallacious statistical thinking. The time has come to keep that promise. Such elaboration seems advisable because I am keenly aware that not everyone will approve of my emphasis upon faulty rather than valid statistical reasoning.

Prior to acquiring real statistical sophistication, most of us pass through two distinct stages in our attitudes toward statistical evidence. Early in life we tend to accept statistical conclusions uncritically on the assumption that figures don't lie. Often we even wilt upon mere exposure to statements beginning with "according to statistics. . ." or "statistics prove. . . ."

As we grow older, however, we tend to swing over to the opposite extreme. We have been too often deceived by advertisers, politicians, prominent citizens with various kinds of axes to grind, journalists seeking to increase the dramatic impact of a point, and on and on—in general, people who have developed considerable skill in drawing faulty conclusions from perhaps flawless data. We find ourselves believing that statistics can prove anything, and therefore can really prove nothing at all. Whereas we once believed that figures can't lie, we now conclude that figures can do nothing but lie. We come to share with Stephen Leacock the cynical notion that "In earlier times they had no statistics, and so they had to fall back on lies. Hence the huge exaggerations of primitive literature—giants, or miracles, or wonders! They did it with lies and we do it with statistics; but it is all the same." Some critics will undoubtedly accuse me of trying to swing you completely toward this attitude of bald skepticism. Such is definitely not my intention as is attested to by the fact that I am an enthusiastic user and teacher of statistics and have been so for quite a few years now.

Libraries are awash with excellent textbooks showing you how to use statistics the right way. I urge you to expose yourself to several of these. Unfortunately, relatively little has been written on the misuses of statistics, and as a result, a potentially worthwhile approach to teaching this broad and fascinating subject has been all but overlooked.

In this book I simply offer a different slant on the subject of statistics in the hope that by studying examples of how things shouldn't have been done you will not only be entertained but also find your judgment sharpened and your ability to appreciate good statistics enhanced. Also, I hope to lead you, as painlessly as possible, toward an understanding of why statistical tools must be used in conjunction with a near-fanatic love for truth.

My own views are quite accurately summarized by Ernst Wagemann:

> We share with Socrates the pious hope that men avoid mistakes once they are aware of them. But we are frivolous enough to suppose that men do this out of a spirit of pure contrariness, and hence are more affected by the sight of a horrible example than a good precept."[7]

---

[7] *A Fool's View of Statistics: The Outline of a Statistical View of the World* (Bern: A. Francke Ag. Verlag, 1950.) This translation from the German appears in W. Allen Wallis and Harry V. Roberts, *Statistics: A New Approach* (Glencoe, Illinois: The Free Press, 1956), p. 65.

2

Some Basic
Measurement and
Definition Problems

When you can measure what you are speaking about and express it in numbers you know something about it; but when you cannot express it in numbers, your knowledge is of a meagre and unsatisfactory kind.
—LORD KELVIN

In either of its two meanings, statistics is intimately tied in with the problem of measurement—*the use of numbers to represent properties.* Before one can study something scientifically, he must be able to express it in numbers, for only then can he distinguish easily and minutely between different but similar properties.

Unfortunately, the ideal way of expressing a property as a number may not be self-evident. Or if it is self-evident, the physical procedures required may be prohibitively expensive or in some other way impracticable. Researchers and professional data gatherers, therefore, must often resort to second- or third-best ways of measuring whatever interests them. The procedures adopted determine to a considerable degree the validity of the figures and the precise manner in which we, as consumers and critics of statistics, interpret them.

In this chapter we consider several kinds of problems related to the task of measuring things. Because it is impossible to measure something meaningfully without knowing what that something is, we must begin by concentrating on the crucial subject of definitions.

The All-Important Definition

Sometimes the task of measuring a property is quite simple. Determining the weight of a sack of potatoes, for example, is for most of us a task of less than staggering proportions. If someone—let us say a philosophy student—were to ask between long, thoughtful draws on his favorite briar, "Exactly how are you defining 'sack of potatoes'?" or "Are you planning to use avoirdupois weight or troy weight?" or "Can you prove the scales are accurate?" we might be moved to respond with something abrasive to a philosophy student's finer sensibilities.

On the other hand, when the thing being measured is "unemployment," "poverty," "marital compatibility," "mental health," "political popularity," or some other concept lending itself to many interpretations, the kind of cautious inquisitiveness displayed by our hypothetical philosophy student becomes absolutely essential. What could be less informative than a collection of figures purporting to measure, say, unemployment, when we are not even certain what kinds of people are counted as unemployed? Are nonworking children included in the count? Are housewives? Professionals? Part-time employees? People on temporary layoff? In most cases, it probably matters less how such problems are handled—provided, of course, that they are handled wisely—than whether we are told or left to guess about which categories of people are counted among the unemployed and which are not.

Whenever a term can be defined in a variety of ways, the data gatherer must decide which of the possible definitions seems most sensible, and, often just as important, which definition lends itself best to efficient, relatively inexpensive data collection. As a result, the definitions used are usually what I shall call *Friendly Definitions*, a term meaning that one of several contending definitions has been settled upon and the data user is asked to accept that specific definition when interpreting the figures. In return for this acceptance, the supplier promises to adhere rigorously to that definition.

Unfortunately, some purveyors of statistical information do not define the concepts they are allegedly measuring, a fact suggesting that our task is primarily one of learning to separate the statistical offerings of suppliers who play by the rules from those who do not. Granted, we must also pass on the adequacy of the definition offered, when one is offered; but when asked to go along with a clearly stated Friendly Definition, chances are we will usually oblige.

Here is a good example of the great importance a clear definition can have in giving meaning to a measurement, in this case a count:

> Whether the city with the world's greatest population is New York or London depends on what areas are referred to by "New York" and "London." The city of London proper had a population in 1955 of only about 5,200, and New York County, or Manhattan, one of the five boroughs of New York City, had 1,910,000. The analogous political units, however, are the City of New York, with a population of 8,050,000 in 1955, and the county of London, 3,325,000 in 1955. Each of these is a municipality made up of boroughs, 29 in London and 5 in New York. A comparison often made (though inaccurately) is that between greater London and the City of New York—probably because of the coincidence that the City of New York, when it was formed by the consolidation of New York, Brooklyn, and other areas in 1898, was referred to as "Greater New York." "Greater London," with a 1955 population of 8,315,000, is defined as the area within 15 miles of the center of the City of London.
>
> It has been estimated that the area within 15 miles of the center of New York has a population of 10,350,000. The "New York Standard Metropolitan Area," however, had a 1955 population of 13,630,000. (A Standard Metropolitan Area is defined by the U.S. Bureau of the Census as a county or group of counties containing at least one city of 50,000 or more, plus such contiguous counties as are metropolitan in character and integrated with the central city by certain specified criteria.) A metroplitan area defined for London on a basis similar to that used for New York have a population of approximately 10,000,000.[1]

The figures are admittedly dated, but the moral isn't. Imagine how meaningless a comparison between the populations of London and New York City would be if the geographic boundaries used were not clearly spelled out. The

[1] W. Allen Wallis and Harry V. Roberts, *Statistics: A New Approach* (Glencoe, Illinois: The Free Press, 1956), p. 68.

necessity of using Friendly Definitions is also demonstrated in this example, for whatever definitions were decided upon, they would necessarily be Friendly according to my Friendly Definition of this term.

Additional examples of the strategic role of definitions are easy to find. In cost studies, for instance, confusion sometimes occurs between the economist's and the accountant's definition of "overhead cost." In economic analysis, overhead costs do not change with changes in the volume of production, whereas accountants sometimes allocate these costs among different years or different products in proportion to the volume of production. Either practice can be justified, but the definition used in a specific instance should be explicitly stated.

The familar and innocuous-sounding word "industry" can plague business economists whether they are charged with measuring the degree of monopoly in an "industry" or studying the extent to which a specific "industry" utilizes the output of other "industries." The measuring devices used are constructed so that the statistical results are totally determined by the manner in which one "industry" is distinguished from others. Any way such a distinction is made is bound to prove imperfect. Is there such a thing as an "automobile-tire industry," for example, or are automobile tires manufactured by the "rubber industry" or some other industry?

Examples of the Statistical Leverage of a Definition

If terms like "overhead cost" and "industry" are trouble makers, "poverty" is a hardened criminal. Efforts to measure the extent of poverty in the United States have been both numerous and well-publicized in recent years. The usual procedure is to (1) designate some level of family income that will serve to distinguish poverty families from the others, and (2) calculate the number of families whose income falls below the specified cutoff point. Despite some shortcomings, which I'll touch on momentarily, this basic approach is probably as good as any other reasonably simple one. At least the figures are quite easy to interpret, a condition which might not exist if the concept were too greatly refined. It should be remembered, however, that the level of family income that spells poverty completely determines the number of families receiving that designation and is necessarily always picked somewhat arbitrarily.

While not wishing to make light of the serious human and social problems implied by the poverty statistics, I must admit to having been grimly amused by the poverty numbers game that some politically oriented economists have been playing, particularly since 1964. To play the game, all one really needs, it seems, is a plausible Friendly Definition of "poverty" that differs somewhat from the definitions used by the other contestants.

The game was being played prior to 1964, but its visibility was increased greatly in that year with the release of a report prepared by the President's Council of Economic Advisors and entitled "The Problem of Poverty in America."[2] The CEA decided that households with incomes of less than $3,000 per annum lived in poverty. Using this income criterion, they calculated that 20 percent of all U.S. households containing some 30 million persons fell into this poverty class. How was the $3,000 figure arrived at? In brief summary, it was obtained by starting with a concept of minimum nutritional adequacy, consisting of specified amounts of calories, proteins, vitamins, etc., and translating these requirements into food items that would do the job. It was determined that, with 1964 prices, these food needs could be supplied for about $5.00 per week per person. For an urban family of four, the diet comes to about $1,000 a year. On the assumption that low-income families should spend about one-third of their incomes on food, a poverty standard of $3,000 was arrived at.

As previously mentioned, any income criterion of poverty is necessarily arbitrary and woefully inadequate in a number of ways, a fact that the CEA was more keenly aware of than anybody. Here are a few of the more apparent inadequacies of the one used: First, it does not take assets into consideration or count as income such things as money from sale of property, borrowed funds, gifts, lump-sum inheritences or insurance payments. Hence, a retired couple who own their own home and enjoy a modest income from stocks and bonds and who, of course, are free to cash in some of their assets at any time they choose, might be classified as a poverty family even though their subsistence needs are well taken care of.

Second, the $3,000 figure is itself, to quite an extent, a reflection of American affluence rather than an indication of the bare minimum income requirement for keeping body and soul together. About one-third of the poverty families had homes and one-half had cars. Most had telephone service and various household durables that might have seemed like extravagant luxuries to our great grandfathers had they been available at all.

Third, transitory poverty is not distinguished from permanent poverty. Temporary unemployment of the main breadwinner due to a sluggish national economy could make the family eligible for inclusion in the poverty count one year but not the next. Maybe that's the way it should be, but we must keep this transitory component in mind when interpreting the figures. Most of us, I believe, think of poverty as a more or less chronic condition.

These disadvantages of the $3,000 poverty criterion, as well as others that might be mentioned, are certainly not news to the Council of Economic Advisors. Nor are they news to the Social Security Administration. (The SSA has recently refined the concept of poverty by getting inflation into the

[2] Presented in *Economic Report of the President* (Washington, D.C.: United States Government Printing Office, 1964), pp. 55–84.

calculations and by taking account of family size and composition, the age of members, and whether the residence is farm or nonfarm.) Despite its several inadequacies, the $3,000 figure for 1964 was probably as good as any reasonable alternative figure that might have been arrived at.

The official count of 30 million persons living in poverty in 1964, however, rather than settling the matter, served to fire the imaginations of other poverty researchers who proceeded to see who could give the count the biggest boost. The poverty numbers game grew quite exciting for a time with several contestants placing the figure in the 40 to 50 million range. But the winner, the one whose definition of poverty exerted the most *Statistical Leverage* (a term I like to use to indicate the extent to which the count changes, given a small change in the definition), seems to be the zealous fellow who counted a startling 80 million poverty-stricken people, a figure representing more than a third of all the people living in the United States.

The game seems to have settled down a little in recent years. Still, each time a new report is issued, a greater commotion follows than is justified by the quality of the figures. According to the SSA report for 1970, for example, the number of people living below the poverty line was 25.5 million, some 1.2 million (5.1 percent) above 1969. This news was widely interpreted as an indication that as a nation we are losing the war against poverty. Is this conclusion justified? Maybe. But let us not forget that 1970 was a recession year and the transitory component, mentioned above, was undoubtedly much inflated.

How many people would you guess were unemployed during the Great Depression of the 1930's? Well, picking out a specific month, we find that in November of 1935 the figure was around nine million. Or 11 million. Or 14 million. Or 17 million. It all depends on whose figures you like. The following bewildering and contradictory list of unemployment estimates all pertain to this one month and all were prepared by reputable agencies:

Table 1. Estimates of Unemployment for the Month
of November 1935 According to Five Reporting
Agencies

Agency Preparing Estimate	Estimate of Number Unemployed
The National Industrial Conference Board	9,177,000
Government Committee on Economic Security	10,913,000
The American Federation of Labor	10,077,000
National Research League	14,173,000
Labor Research Association	17,029,000

Source: Jerome B. Cohen, "The Misuse of Statistics," *Journal of the American Statistical Association.* XXXIII, No. 204, (1938), 657.

To make matters worse, just six months later, as the United States Chamber of Commerce issued a report estimating the number unemployed at 4 million, the *New York Sun* announced that on the basis of a survey of 30 million workers, unemployment amounted to between 3 and 3.5 million. The Labor Research Association insisted that all estimates lower than its own were erroneous. The Chamber of Commerce held that all estimates higher than its own were inaccurate.[3]

These estimates differ, of course, primarily because of differences in the definitions of unemployment used by the various sources. Some estimates took into account unemployment among farm labor and some did not; some included estimates of people leaving school and seeking employment for the first time and some did not; some considered unemployment among professionals and some did not. And so it goes—the considerable variation among the estimates testifying to the sometimes substantial Statistical Leverage exerted by a difference in definition.

As far as this country is concerned, such differences of opinion have been eliminated—officially, at least. The Bureau of Labor Statistics releases unemployment statistics each month that are the most all-embracing in the world. The figures include people who have (1) voluntarily quit their jobs, (2) been discharged for misconduct or poor work, (3) recently found jobs but have not yet reported for work, (4) are available only for part-time or temporary work, (5) simply haven't bothered to look for work because they don't believe that any is available, or (6) are on temporary layoff due to a strike affecting a major customer or supplier of their employer. So just about any age-eligible person who can possibly be construed as unemployed is included in the count. As long as we are aware that the Friendly Definition of unemployment in this country is so all-embracing, we can use the figures in many meaningful ways. Problems still arise, however, when making comparisons between countries. Most foreign unemployment estimates are based on registrations at employment exchanges, a method that results in a relatively lower count than does this country's approach.

Recently, members of a Senate Small Business subcommittee had difficulty determining just how many franchise operations exist in the United States. Robert M. Dias, president of the National Association of Franchised Businessmen, told the Senators there are 1,200 franchisers with 670,000 franchisees doing a total business of $100 billion. John V. Buffington, general counsel of the Federal Trade Commission, citing latest Commerce Department figures, put the franchisers at 1,100 and the franchisees at 400,000. Thomas H. Murphy, publisher of the *Continental Franchise Review*, said there are "con-

---

[3] Cited in Jerome B. Cohen, "The Misuse of Statistics," *Journal of The American Statistical Association*. XXXIII, No. 204, (1938), 657.

servatively" 500,000 franchisees doing a total business of $90 billion.[4] The source of the trouble was, again, lack of agreement on what should be included in the count. Some witnesses construed "franchise business" to be limited to small hamburger and fried chicken, fast-service operations, while others viewed it as including automobile dealers and service stations.

Important Omissions in the Data

Needless to say, once a definition has been decided upon all relevant subcategories should be fairly represented in the reported data. Omission or under-representation of a strategic component can lead to annoying errors in interpretation.

Between 1950 and 1956, for example, some 1.5 to 2 million houses, worth maybe $15 billion, were lost in this country. They weren't lost through fires, floods, or other natural disasters; they were simply misplaced, so to speak, by government statisticians. Instead of 8 million houses going up in those years, as was originally reported, the correct figure was probably almost 2 million (25 percent) higher than that. Recognition of this underestimate led to a program of aerial data gathering over specific counties where reported data on housing starts and value of construction put in place were considered by government statisticians to be especially inadequate.[5]

A more recent example of the potential treachery of an excluded or under-represented component was a $7-billion error made by the Federal Reserve Board when reporting money supply figures for October 1970. These figures are watched closely by analysts of business conditions because changes in the rate of change in the money supply carry implications about future economic growth, inflation, and, in turn, forthcoming Federal Reserve policy decisions. The error resulted from a failure to include a large volume of international dealings in *Euro-dollars*—U.S. dollars held by foreigners. When the figure of $206 billion was boosted to $213 billion, much colorful debate arose around the country regarding the likelihood of an imminent reversal in Federal Reserve policy.[6]

Unfortunately, there is little that even a conscientious user of statistical data can do about recognizing such omissions. And there is absolutely nothing he can do about correcting for them himself. If he is wise, however, he will develop the habit of viewing most highly aggregated national economic data as a tentative mix of fact, estimate, and judgment. A variety of problems related to definitions and methods of data collection still remain and few such problems will be resolved in the very near-term future.

---

[4] *Business Week*, January 31, 1970, p. 30.

[5] *The Wall Street Journal*, November 2, 1959.

[6] *The Wall Street Journal*, November 30, 1970.

Spurious Accuracy

The story is told about a man who, when asked the age of a certain river, replied that it was 3,000,004 years old. When asked how he could give such accurate information, his answer was that four years ago the river's age was given as three million years.

Clearly, the man in this story was unaware that the three-million figure was a crude estimate rather than a precisely known fact. His tacking on the four years was not only unnecessary but potentially misleading as well, for it gave the impression that a degree of accuracy had been achieved that was really unattainable. This is an example of *spurious accuracy*. Many things simply cannot be measured with as much accuracy as some purveyors of statistical information like to pretend.

An automobile advertisement caught the reader's attention with the assertion that on a certain day an estimated 262,825,033.74 tons of snow fell upon the United States.

The official publication of the Austrian Finance Administration stated that the population of Salzburg Province in 1951 was 327,232 people— 4.719303 percent of the entire population of Austria.

A large distillery declared that over the years the company had squeezed 191,752 oranges, 580,582 lemons, and 453,015 limes to make its whisky sours, daiquiris, and margaritas.

The New York Times reported that the St. Patrick's Day parade cost the city \$85,559.61 whereas the Puerto Rican Day parade cost only \$74,169.44.

The Automobile Manufacturers Association reported in *1971 Automobile Facts and Figures* that employment in automotive parts production by companies outside the automobile industry proper was: for narrow fabrics, 1,127 people; for apparel findings and related products, 20,196 people; for points and allied products, 4,959 people; for hardware, 34,303 people; and the list goes on. All told, 37 categories of automotive parts are listed with the related number of employees shown with seeming accuracy right down to the last man or woman. To the AMA's credit was the explanation in an attached footnote revealing that the figures were estimates subject to error.

Examples could be multiplied indefinitely. Seldom in statistical work of any kind can precise figures like those in the above examples be obtained. But the appearance of accuracy suggests to many trusting readers that the source "really knows what he's talking about."

This fallacy is found in all branches of statistical investigation. Whatever the context, one is always wise to be skeptical of figures pretending great accuracy. Usually the simple test of asking yourself "Judging from what I know about the thing being measured, can I really believe that such accuracy is possible?" will be sufficient to keep you from blindly accepting data pretending unattainably high orders of accuracy.

Valid Measures Used Inappropriately

Sometimes a perfectly good measure is used as an imperfect proxy for something else. The following advertisement, for example, is hypothetical but based loosely upon an actual one that appeared in many magazines around the country:

> Smythe's Elixer corrects a variety of scalp diseases and stops the hair loss they cause. Smythe's has been used by over half-a-million people on our famous Double-Your-Money-Back-Guarantee. Only 1% of those men and women were not helped by Smythe's and asked for a refund. This is truly an amazing performance.

The most obviously misleading part of this advertisement is the use of requests for refunds as a measure of the number of customers not helped by the product. What do you suppose is the ratio of dissatisfied customers to the number of customers who actually take advantage of a money-back guarantee for a relatively inexpensive product? One-to-one as implied by the ad? Two-to-one? Ten-to-one? One hundred-to-one? The true ratio, of course, is unknown and quite likely unknowable. Almost certainly, however, using the number of customers requesting a refund to measure the total number of customers who weren't helped by the product leads to a too-low estimate of the truth.

Who has not heard the oft-repeated advertising claim that nine out of ten doctors recommend the ingredients of a certain patent medicine? Let us accept for now the part about the nine out of ten (presumably, though not necessarily, every ten) doctors recommending the ingredients in this product; it does not necessarily follow that they recommend the product by brand name. I am informed by a doctor friend that when the company researchers conduct their marvelous surveys they indicate only ingredients, not brand name, and these ingredients are common to many commercial medicines. This advertising claim appears to be an earnest attempt to mislead consumers into thinking that the product is recommended more often by name than is actually the case.

Automobile registration figures are frequently used to measure the number of automobiles in the hands of the public. However, such figures are not entirely satisfactory for this purpose for at least three reasons: (1) Some states issue a new registration upon sale of a car while some transfer the old registration to the seller's new car; (2) Station wagons, taxis, and some other types of automobiles are classified as passenger cars in some states and not in others; and (3) Some cars are registered to dealers before they are sold to customers.

The use of proxy measures is often justifiable because the thing of interest doesn't lend itself to accurate and/or relatively inexpensive measurement. However, one should at least be able to recognize when a proxy measure is being used and be prepared to pass judgment on its quality. Some proxies are better than others and many are decidedly bad.

In the next chapter we focus more closely on problems of measurement and definition when we consider the rather colorful, though totally unenlightening, subject of *Meaningless Statistics*.

3

Meaningless Statistics

> *. . . a tale told by an idiot, full of sound and fury, signifying nothing.*
>
> —SHAKESPEARE

My favorite Meaningless Statistic is this one attributed to humorist Robert Benchley: "It is not generally known, I believe, that one comic editor dies every 18 minutes, or, at any rate, feels simply awful." It seems to me that if someone is going to waste my time and insult my intelligence by foisting Meaningless Statistics upon me, he should at least have the common decency to see that his offerings are funny. So far, Benchley is the only one to meet my rigorous requirements. Most Meaningless Statistics aren't even very funny. In this chapter we deal briefly with these little time wasters.

Some Typical Examples of Meaningless Statistics

The executive of a certain company claimed that about 75 percent of the entire organization had been with the company for many, many years.[1] Is that not a truly impressive record? On second thought, just how impressive

[1] C. I. Daugherty, "How Dedicated People Build Sales," *Specialty Salesman*, August 1968, p. 10.

is it? The answer, of course, depends entirely upon what is meant by "many, many years," a detail this executive saw fit to spare us.

The organization referred to is a manufacturer and distributor of stainless steel cookware, one of several brands sold exclusively by direct salesmen. I admittedly have no idea what the salesman-turnover rate is for this particular company, but I do know that in most direct selling it is pretty high—so high, in fact, that "many, many years" could conceivably mean four or three or even fewer. (Not that it necessarily does, mind you. We simply have no way of telling from the information given.) The 75-percent figure sounds precise enough, but the entire claim fails to deliver the precision promised because the rest of it is so vague.

This is a fairly typical example of a Meaningless Statistic. A *Meaningless Statistic* is a precise figure used in conjunction with a term sufficiently vague that a Friendly Definition is sorely needed to endow the figure with meaning. But such a definition is not provided, or, if one is provided, it is itself so vague that it doesn't really help.

The term "Meaningless Statistic" was coined by Daniel Seligman in a delightful article called "We're Drowning in Phony Statistics."[2] Seligman cites, among many other examples, the following assertion made by a former U.S. Attorney General: "Ninety percent of the major racketeers would be out of business before the end of the year if the ordinary citizen, the businessman, the union official, and the public authority stood up to be counted and refused to be corrupted." That the underlying thought is plausible as well as praiseworthy is beyond dispute. Nevertheless, in the absence of either generally accepted or Friendly Definitions of terms like "major racketeer" and "stood up to be counted," the 90-percent figure used to dress up the argument is meaningless.

An author asserted that he had studied the food intake of more than 50,000 men and women and found, to his own astonishment, that in one group of 4,500 cases, 83 percent were found to be overweight while undereating. Moreover, only 17 percent were found to be overweight because they over-eat. These facts might have seemed astonishing to the reader as well as to the writer if the latter had only been more clear about what he was calling "overweight," "undereating," and "overeating." If the actual criteria used were no more precise than the description given in this article, then the 17- and 83-percent figures are worthless. Either way, the reader of this particular article is left poorly informed.

An advertisement claimed that 95 percent of key government officials read a certain newspaper but failed to let the reader in on how one goes about distinguishing between a "key official" and one who is less "key."

---

2 *Fortune*, November 1961, pp. 146 ff.

More Subtle Examples

Now and then, we run across statistical information where definitional details are super-abundantly present, but we still find ourselves poorly informed. The following, a statistical tautology of sorts, is one such case:

Toward the end of 1967 considerable publicity was given a figure released by the U.S. Bureau of Labor Statistics to the effect that it costs $9,191 for a family to buy "a moderate living standard." The spurious accuracy of the figure (the amount being shown right down to the last dollar) gives the impression that if anyone *really* knows about "moderate living standards," it is the Bureau of Labor Statistics.

Nevertheless, the Bureau's attempted explanation of the term, presented in a 40-page publication entitled "City Workers' Family Budget," provides a paradigm of circular thinking. First come the qualifications. We are told that the figure (1) pertains only to families in urban areas, (2) is not exactly current, based, as it is, upon 1966 data, and (3) represents a national average. So far, so good. We could probably live with these qualifications without undue strain.

We next must know what, according to the Bureau of Labor Statistics, is a "family." A "family," we are told, is a group consisting of a man who is 38, a wife (age unspecified), a boy of 13, and a girl of eight. To deal with families not conforming to these specifications, the Bureau provided a so-called "equivalence scale," which describes other kinds of households and tells us how their costs compare with the standard family's. As the editors of *Fortune* remarked in a scorching editorial:

> For example, an adult under thirty-five living with three children requires 88 percent as much as the standard family in order to live moderately. These relationships were established by studying in staggering detail, the spending patterns of different kinds of consumers. Data in "City Worker's Family Budget" show, for example, that husbands in metropolitan areas buy straw hats, on the average, once in twenty years; in nonmetropolitan areas, husbands seem to need a straw hat every six years. Boys in metropolitan areas get 12.24 pairs of socks in a year; in nonmetropolitan areas they get only 10.31 pairs. Perhaps that is all their fathers can afford after buying all those straw hats.[3]

Splendid. But what is a "moderate living standard?" That, after all, is the key to understanding the $9,191 figure. Well, here are some of the things it isn't: It is not a minimum or "subsistence" standard. It is not an average living standard. Nor is it a standard required for a sense of well-being. BLS states: ". . . many families can and do spend less than the total amount specified in

---

[3] "Shadowy Statistics (Contd.)" (Editorial), *Fortune*, December 1967, p. 98.

this budget without feeling deprived and without impairing their health or their ability to contribute constructively to our society." So what, I repeat, *is* a "moderate living standard?" Under the heading "$9,191 = $9,191" the *Fortune* editorial concludes:

> In an effort to deal with all these uncertainties, *Fortune* has closely analyzed the data and made the following findings. To achieve a moderate living standard, an average family of the kind specified required $9,191 in the autumn of 1966—just as the BLS says it did. And what do *we* mean by a "moderate living standard?" Why, we mean exactly what BLS means but is, apparently, too embarrassed to blurt out: We mean a living standard that could have been bought by the family for $9,191.[4]

We find, after all is said and done, that knowing that it costs $9,191 for a moderate living standard is to know nothing at all. Our store of knowledge has been expanded by about as much as it would be if someone were to tell us, "All red apples are red." (Perhaps a better analogy would be: "All red apples are rubicund.")

Another kind of Meaningless Statistic is the technical measure that is made to sound like a highly advanced tool but is not explained. An article about life in the Soviet Union, for example, shows a table of so-called "status coef-

[4] *Ibid.*

ficients."[5] From this table we learn that physicists have a status coefficient of 7.64 whereas pilots only achieve 7.62. Radio mechanics are also 7.62, ahead of mathematicians who only merit 7.34 and well ahead of geologists with their mere 7.22.

As readers, we are assured that these status coefficients permit government heads ready information for determining how easy or difficult it will be to provide manpower in a given occupation. Unfortunately, we are told nothing about procedures used in computing these coefficients. Granted, understanding them might require such a high order of sophistication that our inferior minds might not be up to the challenge. But if that is so, why are they shown at all? The simple ranks of 1, 2, 3, . . . would be every bit as informative as status coefficients devoid of clarifying details.

In the next chapter we consider statistics whose meanings are clear but which were arrived at by highly questionable procedures. These I call *Far-Fetched Estimates*.

[5] Vladimir Shubkin, "The Occupational Pyramid: Low and High Status Jobs," *Soviet Life*, September 1971, p. 21.

4

Far-Fetched
Estimates

It ain't so much the things we don't know that gets us in trouble. It's the things we know that ain't so.

—ARTEMUS WARD

Much misleading statistical information comes to us as estimates—educated guesses (the level of educational attainment varying over a stupendous range)—about unknown statistical facts.[1] By their very nature, estimates are approximations; as such, they are almost always wrong. The probable wrongness of estimates, however, is not the particular statistical dragon I hope to slay in this chapter.

Despite the strong likelihood that they will differ from the truth somewhat, estimates often provide useful guides to sensible decision making. The presence of error does not make a statistical fallacy. Unfortunately, our news media present many estimates not worth the eyestrain incurred from reading them. In this chapter we survey five families of *Far-Fetched Estimates*: (1) Unknowable Statistics, (2) Estimates Based on Eccentric Theories, (3) Preposterous Estimates, (4) Build-Ups From Dubious Clusters, and (5) Uncritical Projections of Trends. I also offer a few guides on how to scrutinize an estimate.

[1] When I speak of "estimates" in this chapter I do not, in general, refer to formal statistical estimation of some characteristic of a population. Estimation in this sense, a branch of statistical inference, is touched on in Chapter 12.

Unknowable Statistics

The term *Unknowable Statistic* is another one coined by Daniel Seligman, the author mentioned in connection with Meaningless Statistics in the preceding chapter.[2] According to Seligman, an Unknowable Statistic is one whose meaning is perfectly clear, but the alleged fact is one that no one could possibly know. The figure might be someone's wild guess, a loosely prepared estimate, or maybe even an estimate prepared with painstaking (but wasted) care, but we are not usually told which. If we are told, we aren't really helped because the thing of interest defies accurate measurement or enumeration anyway.

Frequently a statistic is unknowable because some physical barrier interferes with data collection. Sometimes the desired information is unknowable because the only people who might know don't know (or won't say), or because relevant records are woefully inadequate.

Statistics Unknowable Because Some Physical Barrier
Interferes with Data Collection

In his classic article, Seligman dwells at some length on his efforts to learn the origin of statistics on the rat population of New York City. Rat figures had been cited in scattered newspaper and magazine articles as being in the range of 8 million to 9 million, but Seligman found no basis in fact for these figures. He learned that the only real study ever done on the matter was conducted back in 1949. Researchers working under a Rockefeller Foundation grant had actually gone out and counted rats in a certain area. Extrapolating their findings to the entire city, they estimated the number of rats to be no more than about 250,000. The much higher figures presented by some journalists were, apparently, pure fabrications.

Probably one of the most uncontroversial statements of recent years is this observation regarding the difficulty of counting rats made by an authority on the subject whom Seligman interviewed: "You can count a rat on the eighth floor, and then another on the seventh floor, and then another on the sixth—but, after all, you may just be seeing the same rat three times."

Statistics Unknowable Because People Surveyed
Don't Know or Won't Say

Statistics are sometimes unknowable because the relevant information can be obtained only through some kind of survey approach (direct observa-

[2] "We're Drowning In Phony Statistics," *Fortune*, November 1961, pp. 146 ff.

tion being out of the question), but the people questioned either do not know the answers or are reluctant to give accurate answers. Perhaps the questions require the respondent to remember things that he is not in the habit of keeping track of—as, for example, "How many cups of coffee have you consumed during the past thirty days?" Or perhaps the questions deal with illegal or potentially embarrassing activities.

I recently heard a speaker declare: "There are at least five to seven million couples who swap wives in this country in one way or another, sometimes at parties especially for the purpose, and at other times in small groups or sometimes just a pair of couples." The difficulty of obtaining accurate information on wife swapping, especially since only a pair of couples is often involved, is too obvious to merit comment. When the speaker was asked where he got his statistics, he answered that it was "common knowledge among sociologists and a conservative estimate." Such "common knowledge" is at worst misleading and at best pointless and silly.

One of the most unsettling Unknowable Statistics I have run across illustrating the point that sometimes the only people who might know don't know was turned up in a poll, conducted by a university student newspaper, of marital status among freshmen. The results: Single– 1,568; Married– 16; Undecided– 11!

Statistics Unknowable Because Not All Cases
Get Reported

Statistics are frequently unknowable because not all cases get reported or observed. Figures on such activities as abortion, embezzlement, swindles, illegitimate births, child beatings, dog bites, rat bites, shoplifting, etc. are notoriously suspect because many cases simply never get reported.

A congresswoman stated: "Rats have killed more humans than all the generals in history."[3] She is probably right, if only because rats and humans have coexisted none too amicably for many thousands of years whereas the military label "general" is of more recent origin. Furthermore, generals do relatively little direct killing. Substantiating this claim statistically, however, would be hopelessly impossible (as the congresswoman apparently realized, for she made no attempt to do so) because of inadequate reporting, down through the years, of deaths from the two sources.

"Of the contributions made by individuals in this country, it is estimated that $100 million a year goes to phonies—swindlers soliciting for fake charities or unethical professional fund raisers who absorb much of your gift."[4] By their very nature, charity swindlers are generally those whose unconscionable activities are as yet undiscovered. So the $100-million figure should scarcely be taken as highly reliable.

Estimates Based on Eccentric Theories

Early in 1969 many newspapers ran the following quote from a spokesman for the International Flat Earth Society (most of whose members reside in and around Dover, England, and who really do claim to believe that the earth is shaped like a flapjack): "The sun is only 3,240 miles from the earth and only 32 miles in diameter. Can you imagine the summers we would have if it were 93 million miles away as they would have us believe?"[5]

My curiosity aroused, I attempted to learn how these estimates were developed. I found that the figures, and probably the organization itself, could apparently be traced to a spiritual leader named Wilbur Glenn Volivia, who, beginning in 1905, ruled for 30 years over a religious community of some 6,000 members of the Christian Apostolic Church in Zion, Illinois.

[3] *U.S. News and World Report*, July 31, 1968, p. 10.

[4] Don Wharton "Four Frauds to Beware of This Christmas," *Reader's Digest*, December 1968, p. 93.

[5] This specific quote came from *The Manila Times*, January 3, 1969.

Volivia is said to have offered $5,000 to anyone who could prove that the earth is spherical. Paradoxically enough, he is purported to have made several trips around the world lecturing on the subject. (He never accepted the notion that he had circumnavigated a globe, however, but instead insisted that he had merely traced out a circle on a flat surface.) Here is what Volivia had to say about the sun:

> The idea of a sun millions of miles in diameter and 91,000,000 [sic] miles away is silly. The sun is only 32 miles across and not more than 3,000 miles from the earth. It stands to reason that it must be so. God made the sun to light the earth and therefore must have placed it close to the task it was assigned to do. What would you think of a man who built a house in Zion and put the lamp to light it in Kenosha, Wisconsin?[6]

Although the manner in which the specific estimates were derived is not greatly clarified by this passage, we do learn that the figures are based upon an ultra-literal interpretation of certain biblical passages and a common-sense view of God's ways. In other words, the figures were not merely plucked out of thin air; they do possess an underlying rationale of sorts. Just how the modern-day Flat Earth Society's spokesman learned that the sun is 3,240 miles from the earth rather than only 3,000 or less as Volivia contended, or, for that matter, how Volivia hit upon the 3,000 figure in the first place, is still unknown. What is known is that the figures were somehow distilled from scripture and from the assumption that God would have the good sense to place the sun close to the earth.

This is an example of an *Estimate Based On An Eccentric Theory*. Such

[6] Quoted in Martin Gardner, *Fads and Fallacies in the Name of Science* (New York: Dover Publications, Inc., 1956), p. 17.

estimates are dependent upon faulty assumptions above whose quality they cannot rise. They just about have to be wrong.

Forecasting the state of the economy is one kind of activity where estimators with eccentric theories have always abounded. Joseph (of biblical fame) probably started the ball rolling when he interpreted Pharaoh's dream and came up with a correct prediction of an imminent 14-year economic cycle. Soothsayers since Joseph have studied everything from the positions of heavenly bodies to the gauge of the hemline of ladies' skirts to get an advance clue of the state of business conditions. What follows is a brief summary of some of the more colorful approaches to preparing economic estimates:[7]

1. The Jupiter-Saturn cycle. This forecasting approach depends upon the juxtapositions of the planets. As one zealous advocate explains it, when the planets achieve certain geometric relationships with the sun, they affect the electromagnetic field around the earth and alter man's psychological bearing. "Adverse psychological conditions prevail when two planets are in conjunction on the same straight line from the sun or in squares at a 90-degree angle, with the sun at the center. Favorable conditions exist when the planets form a 60-degree sextile with the sun or a 120-degree trine." Any questions?

2. Levels of Lakes Michigan and Huron. Water levels of Lakes Michigan and Huron are closely watched by some prognosticators on the grounds that, as the water level drops, load capacity falls on ore boats, shipping expenses rise, and steel industry costs go up. Increasing costs in the steel industry can, the advocates argue, affect the regional economy and, indirectly, the nation as a whole.

3. Average size of tabs and tips at Sardi's Restaurant. The owner of this famous New York restaurant which caters to celebrities and celebrity admirers theorizes that business falls off when buyer traffic into New York drops off, a condition quickly reflected in the average sizes of both tabs and tips at Sardi's. "Once the buyers stop coming in, the sellers get hit, and then the economy."

4. Purchases of gourmet foods. Advocates of this forecasting device claim that buyers shift from gourmet foods to more standard fare months before a recession actually sets in.

5. Ups and downs of women's hemlines. The thinking behind the use

---

[7] For this summary I have depended heavily upon "You Can Also Read Tea leaves," *Business Week*, November 17, 1962, pp. 121 ff.

of this indicator seems to be that the distance from the ground at which the hemline is worn reflects relative degrees of optimism or pessimism on the part of women, the people who make most of the consumer purchases. Whether optimism is rising or falling determines which direction the economy will soon be going.

Most of these unusual business indicators have worked well enough in the past to encourage the believers. They have not, however, either individually or collectively, measured up to the exacting requirements of economic indicatordom imposed by the National Bureau of Economic Research, the authoritative New York research organization which passes judgment on many hundreds of would-be economic harbingers.

Preposterous Estimates

Sometimes estimates are wrong on their face. Leonard Engel during the early 1950's collected the following examples of Preposterous Estimates provided by members of the prestigious medical profession:

> An example is the calculation of a well-known urologist that there are eight million cases of cancer of the prostate gland in the United States—which would be enough to provide 1.1 carcinomatous prostate glands for every male in the susceptible age group! Another is a prominent neurologist's estimate

that one American in twelve suffers from migraine; since migraine is responsible for a third of chronic headache cases, this would mean that a quarter of us must suffer from disabling headaches. Still another is the figure of 250,000 often given for the number of multiple sclerosis cases; death data indicate that there can be, happily, no more than thirty or forty thousand cases of this paralytic disease in the country.[8]

Sometimes an estimate will seem perfectly reasonable at first and then crumble under close critical scrutiny. Max Singer, for example, does a brilliant job of talking back to an estimate in an article entitled "The Vitality of Mythical Numbers."[9] Singer points out that it is generally assumed that heroin addicts in New York City steal some two to five billion dollars worth of property in a year's time. Such estimates are obtained, according to Singer, by accepting as fact the frequently cited estimate that there are 100,000 heroin addicts in New York City. These addicts must spend about $30.00 a day on their habit. Therefore, heroin users in New York City must have some $1.1 billion a year to pay for their 365 fixes each (100,000 × 365 × $30.00 equals just under $1.1 billion). Since the property stolen must be disposed of through fences who pay only about one quarter or less of what it is worth, addicts must steal some four to five billion dollars a year just to pay for their heroin.

Even if one makes allowances for the fact that (1) many heroin users who make their livings illegally spend upwards of a quarter of their time in jail, (2) some of what the addict steals is cash, none of which has to go through fences, and (3) a large part of the cost of heroin is paid for by dealing in the heroin business itself and another large part by activities related to prostitution, rather than stealing, one still must conclude, Singer argues in behalf of the estimators he soon refutes, that if there are 100,000 addicts in New York City, they would have to steal at least one billion dollars worth of goods in a year's time.

Approaching the subject from the other side—that is, by developing estimates for the amount of property that is stolen in a year via shoplifting, burglary, and muggings, Singer arrives at a dramatically lower estimate. Using many intentionally high estimates and paying meticulous attention to appropriate details, Singer concludes that heroin addicts cannot be responsible for more than about $330 million worth of stolen property per year.

Singer also questions the validity of the frequently used 100,000 figure for the number of heroin addicts in New York City. Approaching the subject from a variety of angles, he concludes that 70,000 is a more realistic, though probably high, estimate. Singer's own estimates, of course, are hardly paragons of accuracy. Their chief virtue is that they set some reasonable

---

[8] "Danger: Medical Statistics At Work," *Harper's Magazine*, January 1953, p. 81.

[9] *The Public Interest*, No. 23 (Spring 1971), pp. 3–9, © National Affairs, Inc., 1971.

upper boundaries to this problem of theft by heroin addicts. The article is good reading for the statistical critic.

The Build-up from a Dubious Cluster

Another relatively common kind of Far-Fetched Estimate is the *Build-up From A Dubious Cluster*. By "dubious cluster" I refer to a group of people or items undeserving to be used as a basis for broad generalization either because it is unrepresentative of the relevant population (the aggregate of people or items of interest), or because the figure used as a count or measure of the "cluster" is an Unknowable Statistic or is in some other way unreliable. Here are two examples:

A very bad newspaper article described the procedure used by a sheriff's investigator to estimate the amount of money lost each year from shoplifting.[10] He began, the article stated, by making the rounds of 26 supermarkets making up a certain grocery chain and somehow concluded that losses resulting from shoplifting activity for this chain amount to some $30,000 per year. How he arrived at this amount is anybody's guess because it certainly falls into the category of Unknowable Statistics. Then, using this figure as a base and assuming that the particular grocery chain investigated is perfectly representative—both in size of stores and susceptibility to shoplifting—of all grocery chains in the country, he proceeded to estimate losses due to shoplifting for the entire nation by multiplying by the number of food chains in the fifty states. He concluded that shoplifting (for food chains only, I presume, although the article was rather vague on this point) is a million-dollar-a-year racket. (The article was also rather vague on whether the million-dollar figure was supposed to be taken as meaning literally a million dollars or simply as a figure symbolizing the idea that shoplifting is big business.) That this investigator's estimate is worthless despite the empirical approach and some use of arithmetic is self evident.

An article telling direct salesmen how to make big money selling a new brand of faucet washers used a novel chain of reasoning to make faucet washers sound like the most important commodity on the face of the earth.[11] The author asserted that in one city alone (the specific city not being mentioned by name) four to five million gallons of water are lost daily through leaky faucets. Using this Dubious Cluster as a start, he calculated that some 2.5 billion gallons of water are lost yearly in this city. He then contemplated the implications for the nation as a whole and took into account the increasing water needs of a rapidly growing national population. The article fell just

[10] *The Rocky Mountain News* (Denver, Colorado), December 1, 1968.
[11] "Big Money In Leaky Faucets," *Specialty Salesman*, November 1963, pp. 46–8.

short of conveying the impression that before very many years each citizen of this country will be standing in water up to his clavicle if he fails to do his patriotic duty and buy faucet washers.

Uncritical Projection of Trends

As I mentioned in connection with the bizarre economic indicators, some Far-Fetched Estimates concern the future. This is not the place to survey forecasting techniques, some of which are quite elegant and safely out of reach of blanket criticism. However, one of the most commonly used, and abused, forecasting devices does deserve treatment here—namely, the projection of past trends. When used in conjunction with much knowledge about the variables playing upon a series of past numbers, trend fitting and projecting can be most helpful; the use of the tool is not, per se, a statistical fallacy. It is when judgment is abandoned and "arithmancy" allowed to rule that the threat of Far-Fetched Estimates arises.

Sometimes even the *arithmancy* (a derogatory term suggesting that somehow the manipulation of past numbers has imbued the resulting estimates of future numbers with a kind of magic) is pretty sloppy, consisting as it often does of taking X numbers and determining what $X + 1$ would have to be to keep the series growing in *exactly* the same manner as in the past, and then determining what $X + 2$ must be to effect this same end, $X + 3$, $X + 4$, and so

, it goes, further and further into the future. For example, if the United States had a population of 90 million in 1910 and 180 million in 1960, the "arithmanticist" would say that population doubles every 50 years and that we'll have 360 million people in the year 2010, 720 million by 2060, and just under a billion and a half by 2110. "Eventually, using this logic, there will come a day when the radius of human flesh will expand at the speed of light," as one expert testified before a congressional committee.[12]

Other exciting conclusions can be reached by projecting trends unhampered by logical considerations or even common sense. For example: Miniskirts will be shortened to the point where they eventually disappear completely (or, conversely, maxi's will be lengthened to the point that they drag on the

ground for ten miles behind the wearer); The entire earth will be covered with concrete; School teachers will be required to complete thirty years of post-doctoral work; A sprinter will run the 100-yard dash in no time at all—or even finish the race before it begins; and The Lower Mississippi River will one day be a mere mile and three quarters long, as Mark Twain insists in the following much-quoted paragraph from *Life On The Mississippi.*

> In the space of one hundred and seventy-six years the Lower Mississippi has shortened itself two hundred and forty-two miles. That is an average of a

[12] Ben J. Wattenberg with Richard M. Scammon, *This U.S.A.* (New York: Pocket Books, 1967), p. 24. © Doubleday & Company, Inc.

trifle over one mile and a third per year. Therefore, any calm person, who is not blind or idiotic, can see that in the Old Oölitic Silurian Period, just a million years ago next November, the Lower Mississippi River was upward of one million three hundred thousand miles long, and stuck out over the Gulf of Mexico like a fishing-rod. And by the same token any person can see that seven hundred forty-two years from now the Lower Mississippi will be only a mile and three-quarters long, and Cairo and New Orleans will have joined their streets together and be plodding along under a single mayor and a mutual board of aldermen. There is something fascinating about science. One gets such wholesale returns of conjecture out of such a trifling investment of fact.

How to Scrutinize an Estimate

As yet, no resourceful statistician has devised a pat formula for distinguishing between carefully prepared estimates and carelessly thrown-together "guesstimates." Moreover, no set of simple rules is likely to go very far toward helping you make such distinctions yourself. Nevertheless, I would like to suggest that five questions should be asked of any estimate. These questions will serve as a simple screening device for skimming off the worst offenders.

1. What kind of reputation does the source enjoy as a supplier of statistical estimates or as an authority on the relevant subject? Often, estimates of questionable validity are made on a one-shot basis. In such a case, the purveyor of the figures may not be in the business of supplying data and feels under no pressure to establish and maintain a reputation for integrity. Also, frequently he will not be a widely recognized authority on whatever the subject is that the estimates relate to. Certainly, such is case with many examples presented in this book.

2. Does the supplier of the data have an "axe to grind?" Intentional and unintentional bias shows clearly in many of the examples presented in this book. Bias based upon racial or religious bigotry is reflected in some of the examples; bias based upon a specific editorial commitment is reflected in others; bias based upon the desire to sell an idea or a commodity is reflected in still others. Be skeptical of estimates from sources with "axes to grind." Make them work especially hard to convince you that their estimates are reasonably trustworthy.

3. What supportive evidence is offered? Often the answer will be an unqualified "none." When "evidence" is offered, ask yourself whether it does, in fact, support the estimates or whether it has been attached merely for the sake of making a good impression.

4. Does the underlying assumption, theory, or methodology seem

okay? Beware of Estimates Based On Eccentric Theories and well-intentioned but questionable assumptions. Beware particularly of estimates obtained from a Build-Up From A Dubious Cluster and an Uncritical Projection Of Trends. These forms of arithmetic legerdemain are not only common but also sometimes exceedingly subtle.

5. Do the estimates appear plausible? Sometimes common sense and just a little knowledge about the relevant subject is all that is required to strip worthless estimates of their respectable facades.

When you have asked and answered to the best of your ability these five questions and are still not sure whether a particular estimate merits acceptance, you might choose to fall back on the advice of one of the all-time great statisticians, Karl Pearson:

> It can scarcely be questioned that when the truth or falsehood of an event or observation may have important bearings on conduct, over-doubt is more socially valuable than over-credulity.[13]

Let us now turn our attention to *Cheating Charts*.

[13] *Grammar of Science* (London: Adam and Charles Black, 1911), p. 54.

5

Cheating Charts

Get your facts first and then you can distort 'em as much as you please.

—MARK TWAIN

I don't recall exactly when the urge to write a book about statistical fallacies first hit me, but it might well have been about a dozen summers ago when I was a graduate student looking around for part-time work. The prospective job that appealed to me most—until my disenchantment—was that of mutual fund salesman. An important visual aid used by the company's salesmen was a film of still pictures designed to show prospective investors the benefits of buying into this particular mutual fund and the dangers of not doing so immediately.

Among the pictures was a chart of a rapidly rising line depicting assets per share of the company's stock over several recent years. The line was zooming upward at breakneck speed, but, unfortunately, the speed of ascent was given a boost by some tomfoolery with the vertical scale. Cartoons suggestive of the great affluence to be enjoyed by the investor in this fund were placed at intervals along the vertical axis in much the same way as in Figure 5-1 below. Admittedly, the cartoons enhance the entertainment value of the chart, but they also contribute brazenly to the impression that assets per share had been racing upward at a rate far in excess of the true rate.

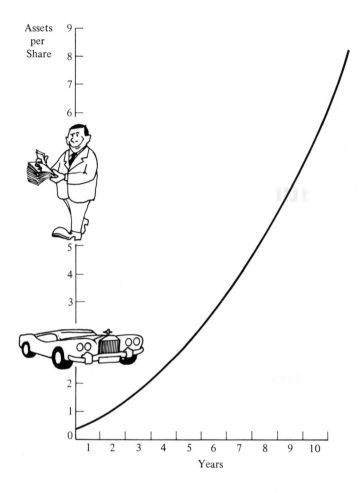

Figure 5–1

In the same film, another chart was used to make the viewer aware of the considerable inflation that had been occurring. It showed the Consumer Price Index, used as the measure of the cost of living, rising rapidly and the purchasing-power-of-the-dollar index declining just as rapidly. The chart looked roughly like the one shown in Figure 5–2.

The unwary viewer tends to interpret the gap between the two lines as the amount of inflation—and it looks like an awfully lot of inflation. Actually, the two indexes measure the same thing, the purchasing-power-of-the-dollar

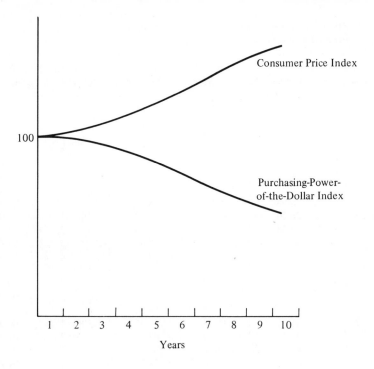

Consumer Price Index

100

Purchasing-Power-
of-the-Dollar Index

1 2 3 4 5 6 7 8 9 10

Years

Figure 5–2

index being essentially the reciprocal of the Consumer Price Index. Only one of the two lines is needed to give a visual indication of the amount of inflation. Use of both lines amounts to double counting and gives the impression of twice as much inflation as had actually been occurring.

I didn't sell for that company.

It is probably safe to say that no statistical tool is used more often to deceive the unwary than the statistical chart. The rest of this chapter is devoted to showing other ways that statistical charts are used to mislead unsuspecting people who really deserve better treatment.

The Line Chart

Table 2 on page 46 shows what is known technically as a bunch of numbers. These particular numbers represent quarterly values of Standard and Poor's index of 500 common stocks for the years 1953 through most of 1971.

Table 2. Standard and Poor's Index of Common Stock
Prices by Quarters, 1953 to 1971

Year and Quarter	Index Number	Year and Quarter	Index Number	Year and Quarter	Index Number
1953–1	26.01	1959–1	55.51	1965–1	86.57
2	24.50	2	57.51	2	87.43
3	23.98	3	58.93	3	86.93
4	24.43	4	57.76	4	91.76
1954–1	26.02	1960–1	56.28	1966–1	91.63
2	28.44	2	56.07	2	88.15
3	30.77	3	55.72	3	81.43
4	33.53	4	55.33	4	79.82
1955–1	36.30	1961–1	62.00	1967–1	87.08
2	38.38	2	65.98	2	91.66
3	43.15	3	66.83	3	94.44
4	44.14	4	70.27	4	94.54
1956–1	45.36	1962–1	69.86	1968–1	91.63
2	46.95	2	62.22	2	98.02
3	48.04	3	57.83	3	99.92
4	46.15	4	59.62	4	105.21
1957–1	44.31	1963–1	65.55	1969–1	100.93
2	46.46	2	69.67	2	101.67
3	46.11	3	70.97	3	94.47
4	40.64	4	73.27	4	94.28
1958–1	41.50	1964–1	77.55	1970–1	88.70
2	43.60	2	80.30	2	79.20
3	47.55	3	82.88	3	78.74
4	52.31	4	84.75	4	86.23
		1971–1	96.73		
		2	101.47		
		3	98.54		

Source: *Business Conditions Digest*, Department of Commerce, Bureau of the Census, July 1970, p. 108 and recent issues.

What do you make of these figures? Unless you already have a trained statistical eye, the numbers probably don't tell you anything except perhaps that the general drift of stock prices has been upward over the years. The difficulty of making sense out of a mass of data is precisely the reason why charts are so often used. Tables filled with figures serve up too much detail for the mind to assimilate. Figure 5-3 shows the same data in the form of a line chart:

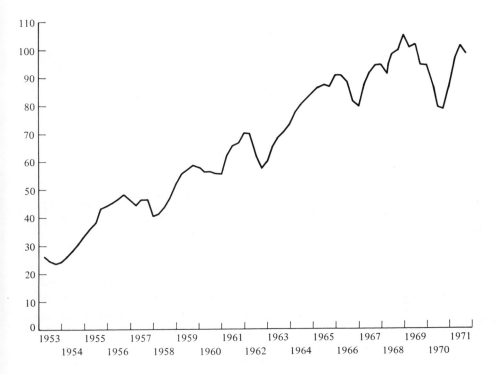

Figure 5–3

In this form, not only is the long-run upward trend of stock prices clearly revealed but so also are several shorter-term cyclical swings and some occasional irregular fluctuations. One gets, at a glance, a pretty good feel for what has happened to stock prices over this particular period.

Figure 5–3 utilizes a so-called arithmetic grid on both the vertical and the horizontal axes. With such a grid, equal numerical changes in the figures show up on the chart as equal-sized vertical movements. Often data shown over relatively long stretches of time are charted in a different manner. The vertical axis will be manipulated (legitimately) in such a way that equal percent changes in the data, rather than equal numerical changes, are seen on the chart as equal-sized vertical movements. Figure 5–4 shows these same numbers charted in this way. The vertical axis has a so-called logarithmic, or ratio, grid rather than an arithmetic grid. That is, the vertical-scale values are spaced in accordance with the differences between their logarithms rather than in accordance with the actual numerical differences. The horizontal axis has a regular arithmetic grid. This kind of chart is called a *semilogarithmic chart*.

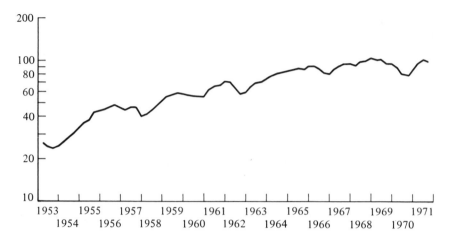

Figure 5-4

In Figure 5-4 recent fluctuations are dampened somewhat as a result of larger base values. That is, a drop of ten index points from a base value of, say, 100 would be only half as deep on the chart as a drop of ten index points from a base value of 50 because the former represents only a 10-percent decline while the latter represents a 20-percent decline. Use of either kind of grid, arithmetic or logarithmic, on the vertical axis is perfectly legitimate; the choice will depend upon whether one wishes to present a picture of absolute or relative changes.

A line chart can be made even more informative if certain other embellishments are used. Figure 5-5, for example, shows our data plotted on the same kind of grid as in Figure 5-4 but with shaded vertical columns appearing here and there. These vertical columns represent periods of contraction (recession, if you prefer) in general business activity. With the help of these shaded columns we see immediately that (1) stock prices have gone through a period of contraction each time the general economy has gone through one, (2) stock prices occasionally experience contractions unaccompanied by contractions in general business activity, and (3) cyclical turning points in stock prices invariably occur somewhat earlier than turning points in general business activity. These tendencies are all but invisible to the eye in Table 2.

The three charts (Figures 5-3 to 5-5) are all pretty good. They are free of broken vertical axes; they have reasonable vertical scales—neither greatly compressed nor unduly stretched out; they cover a long enough period of time that the behavior of stock prices in a variety of business-cycle settings can be seen readily; and no double counting or arbitrary tampering with the scales has been resorted to as was the case in the examples introducing this chapter. Now let us see what options are open to someone with an interest in deceiving people about the behavior of stock prices.

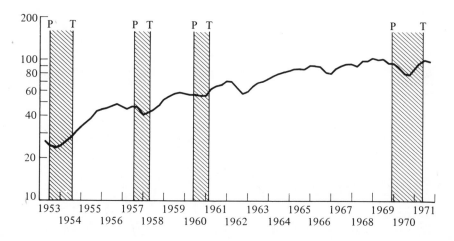

Figure 5–5

A Recent Experience

My wife and I served as an audience of two at a sales presentation given by a representative of a land development company. Among the points made during the presentation was that land is a better investment than any alternative, including stocks. I don't care to argue this point one way or the other. My reason for referring to this experience is that the salesman relied heavily upon a line chart designed to show the behavior of stock prices in the most unfavorable possible light. A summary of the steps taken to achieve this optical illusion should be instructive.

To begin with, the data used in constructing the chart began with 1965. Now there is, of course, no rule that says one can't begin a line chart with whatever year he pleases. But, as Figure 5–6 shows, a beginning year of 1965 makes the course of stock prices seem especially flat. One gets no feel for the long-run upward trend so evident in the charts shown above. Also, an arithmetic rather than a logarithmic grid was used on the vertical axis so that declines showed up in full glory.

To make matters worse, the chart stopped at the third quarter of 1970 even though the sales presentation was given almost a year after that date. This device allowed the chart to show stock prices in the most recent period looking particularly sick. The robust upswing beginning in late 1970 simply wasn't shown. Figure 5–7 shows what happens when this upswing is omitted from the chart.

As if the upward trend in stock prices hadn't already been concealed enough, the creator of this chart saw fit to chop off the bottom of the vertical axis, as shown in Figure 5–8. As a result, not only does the trend of stock

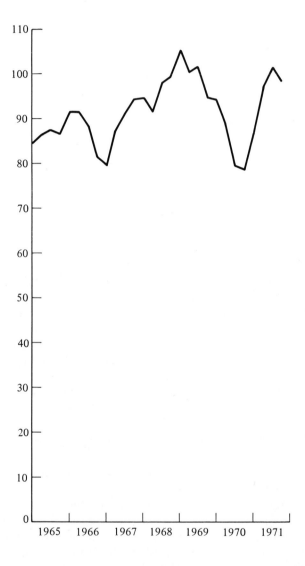

Figure 5–6

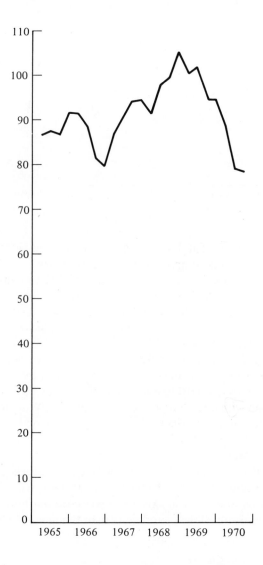

Figure 5–7

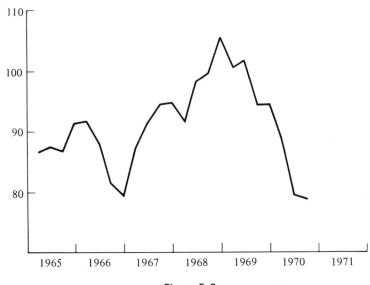

Figure 5–8

prices seem particularly flat, but also the height attained before the period of relative flatness set in is completely withheld from view.

That pretty much exhausts the list of tricks used to construct this particular misleading chart, but there is one more possibility that should be mentioned while we're on the subject. If the jitteriness of stock prices rather than the flatness of the trend were the point being emphasized, the chart could have been constructed with an elongated vertical axis like that shown in Figure 5–9 where each mark up the side represents two index points rather than 10. There is no rule against such tampering with the scales and it does make stock prices seem terribly undependable.

Buy why stop with stretching out the vertical axis? Why not compress the horizontal axis as well? As Figure 5–10 shows, this added trick makes stock prices look about as nervous as one can make them look short of tampering with the actual data.

Of course, if one wanted to convey the idea that stock prices are really quite stable, he could use some the the tricks just discussed in reverse. Figure 5–11 has a compressed vertical axis and a stretched-out horizontal axis. The resulting line looks much less capricious than the one in Figure 5–10.

More on Semilog Charts

We passed rather hurriedly over the advantages that sometimes result from the use of a logarithmic rather than an arithmetic grid on the vertical axis. Some further elaboration seems in order.

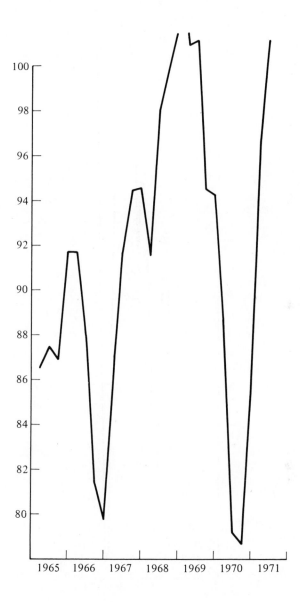

Figure 5–9

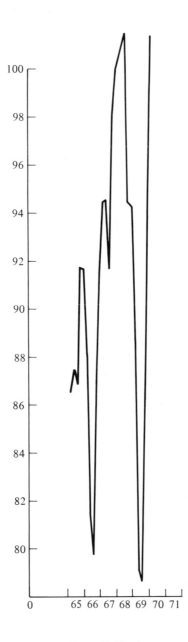

Figure 5–10

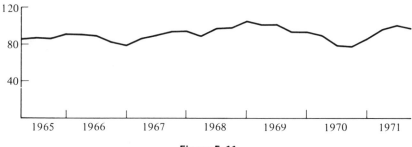

Figure 5–11

Suppose that *Better Houses* magazine has been pre-eminent in its field for a great many years. Suppose further that five years ago, *Rosey Residences* magazine, a direct competitor, appeared on the scene. Finally, suppose that the recent five-year history of number of subscribers to the two magazines has been as shown in Table 3.

Table 3. Number of Subscribers to *Better Houses* and *Rosey Residences* Magazines, 1968 to 1972

Year	*Better Houses* (*millions*)	*Rosey Residences* (*millions*)
1968	10.2	0.4
1969	12.1	0.6
1970	13.8	0.9
1971	15.0	1.3
1972	18.5	2.2

Source: Hypothetical data.

If subscription figures were to be used in advertisements for the two magazines, *Better Houses* would benefit from showing the data on an arithmetic scale. Figure 5–12 conveys the definite impression that *Better Houses* is doing better than its younger competitor not only in terms of absolute number of subscribers but in terms of year-to-year increases as well.

Keep in mind, however, that *Rosey Residences* is only five years old. No one could reasonably expect it to be performing on a scale comparable to its older competitor. Interest, therefore, should not center either on absolute level or absolute increases in subscribers but rather upon relative increases. If *Rosey Residences* were to use these subscription figures in its advertising, it would be well advised to use a logarithmic vertical grid to show that its relative increases have been greater than those of *Better Houses*. The faster relative growth of the younger magazine shows up dramatically when plotted

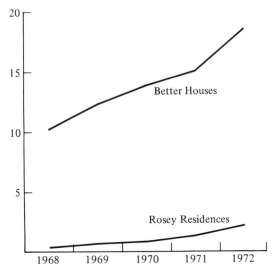

Figure 5–12

in this way, as Figure 5–13 demonstrates. When viewed in this light, *Rosey Residences* is seen to be giving the older magazine a pretty good run for its money.

None of the preceding is meant to suggest that arithmetic grid charts are intrinsically bad nor that semilog charts are intrinsically good. To repeat a point made earlier: The choice of grid will depend entirely upon what one wishes to convey with the chart. In this example, relative increases were assumed to be of paramount importance; therefore, the semilog grid was the more appropriate choice. The mere fact that creators of statistical charts have a choice, however, opens the door to possible abuses. As a statistical critic, you will be wise, when confronted with a chart prepared by someone else, to ask yourself whether the impression conveyed would be materially different if the other kind of grid were used and, if so, whether the other grid might not have told the story better with less distortion of the data.

Unmarked Axes

In view of what has already been said about line charts, it almost goes without saying that charts with unmarked axes should never be trusted. When the creator of a chart is free from the discipline imposed by well-marked axes, he can convey through the chart just about any impression he chooses. In Figure 5–14, for example, we see two sales lines rising by quite different amounts over time. Or do we? When numbers are placed along the

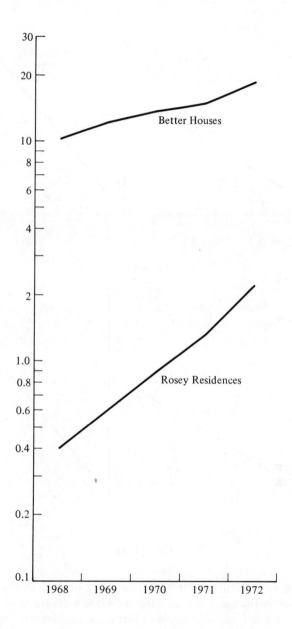

Figure 5–13

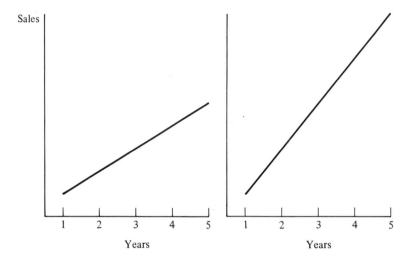

Figure 5–14

Figure 5–15

vertical axis, as in Figure 5–15, it becomes clear that the two sales lines are rising by exactly the same annual amount, 4,000 units per year. The second chart simply has a more stretched-out vertical scale than the first.

Advertisers sometimes come up with charts that are totally meaningless not only because the scales are unmarked but also because other untranslatable embellishments are thrown in for good measure. A testimonial-type ad, for example, might go something like this: "Lesper Knight of Pine Foursome,

Vermont, says, 'My virility was low and sinking lower until I tried Dr. Horney's Hormone Wafers. Thanks to Dr. Horney's Hormone Wafers . . .' (and so forth)." And then a chart is shown which supposedly proves that the testimonial is valid. Ponder Figure 5–16 and see how long it takes you to go out of your mind.

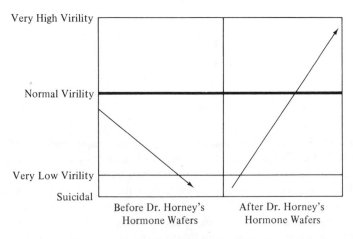

Figure 5–16

This example is hypothetical, of course, but the chart, if it can be called a chart, isn't too different from some that have a way of popping up in advertising every little while. Look upon them as entertainment and I promise you some chuckles. Look upon them as something that can be interpreted with any precision and I promise you an extended course in basket weaving.

Bar Charts and Pictograms

In case the subject ever comes up in polite conversation, about 4.2 times as much money was spent on housing in 1970 than was spent in 1946. The figure for expenditures for residential structures (a term referring essentially to one-family houses and apartment units) in 1946 was a mere $7.2 billion whereas the 1970 figure was $30.4 billion. What does this have to do with statistical charts? It provides us with a pair of figures lending themselves to convenient visual comparison by means of a bar chart. In bar chart form the two figures appear as shown in Figure 5–17.

Just as long as the widths of the bars are equal, as they are in Figure 5–17, the bar chart allows us to compare housing expenditures for the two years simply by eyeballing the lengths of the bars. Of course, the bottoms of the bars shouldn't be chopped off, as they often are, or the visual comparison becomes invalid.

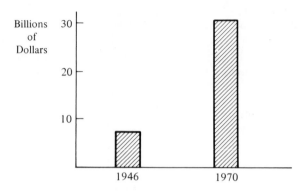

Figure 5–17

The only trouble with bar charts, or so many people seem to believe, is that they are terribly uninteresting things to look at. As a result, creators of charts often use pictures of objects rather than bars to tell their stories. Use of such *pictograms*, as they are called, might indeed enhance the entertainment value of a chart, but it also usually introduces some unavoidable technical problems. For example, if the amount spent on housing in 1946 were indicated by a house looking like that shown in Figure 5-18, how should the amount for 1970 be shown? By 4.2 houses of the same size as in Figure

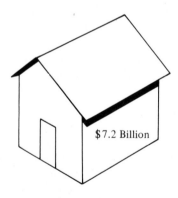

$7.2 Billion

Figure 5–18

5-19? Presenting the figures in this manner has at least two shortcomings. First, the 1970 figure is not shown outright but must be calculated by the viewer. Second, the viewer might easily misconstrue the intention of the pictogram and tuck away the impression that whereas the typical American family owned only one house in 1946, it owned 4.2 houses in 1970—an absurd conclusion to be sure but one easily reached if the chart and surrounding expository material, if any, are not scrutinized carefully.

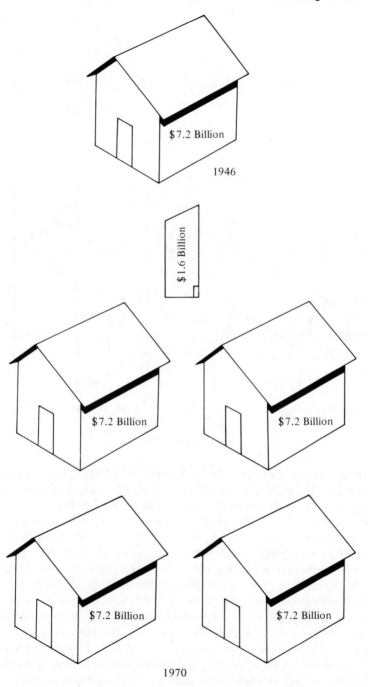

Figure 5–19

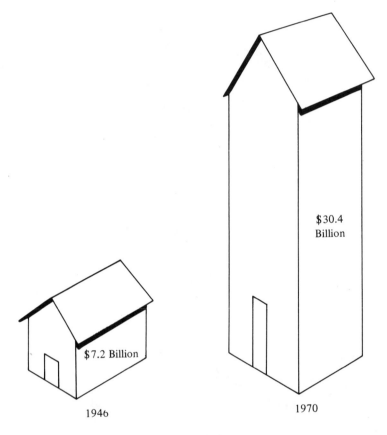

Figure 5–20

An alternative that could be used would be to show the 1946 house as being a certain height and the 1970 house as 4.2 times as tall. But this procedure presents problems of its own, as can be seen in Figure 5–20. Figure 5–20 is accurate enough, but the 1970 house is a pretty silly looking structure. For this reason, the second house might be drawn so that the width is also increased by a factor of 4.2, as has been done in Figure 5–21. Now the second house looks more like a house, but a serious technical error has been introduced. Considering only the two dimensions, height and width, the second house is seen to be larger by a factor of 17.64 rather than only 4.2 (4.2 times 4.2 equals 17.64). But that isn't the whole story either. Because the pictures are of three-dimensional objects, the viewer gets the impression that the total area represented by the second house is over 74 times as great as for the first (4.2 times 4.2 times 4.2 equals 74 plus). Of course, he can look at the numbers shown on the side of each house and, if he reflects for a moment, get the meaning of the chart straight in his mind. But it does require a little reflec-

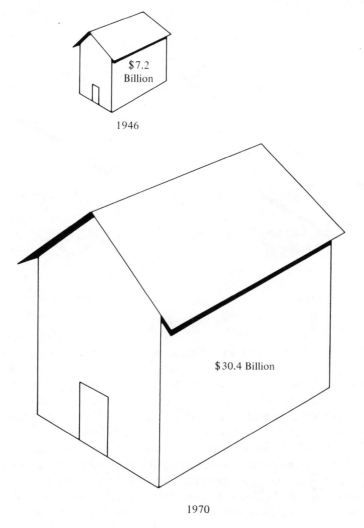

$7.2
Billion

1946

$30.4 Billion

1970

Figure 5–21

tion, a sometimes painful activity, and, in the meantime, a seriously misleading visual impression is conveyed.

To correct the chart so that the area is accurate in two dimensions, the
1970 house should be $\sqrt{4.2}$ equals 2.049 times as high and 2.049 times as wide,
as shown in Figure 5–22.

Unfortunately, two areas cannot be compared with the same ease as can
bars having, in effect, only one linear dimension. But this is a characteristic
of any two dimensional chart and should always be taken into account.
Figure 5–22 tells the story adequately if the viewer will be obliging and view

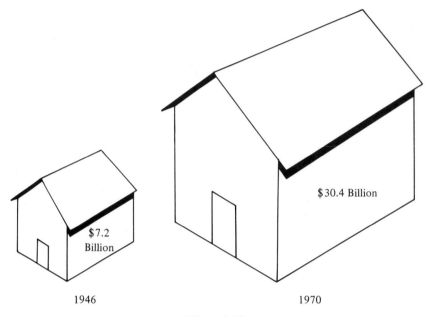

Figure 5–22

the houses as having only two dimensions. Of course, unless he has had considerable experience around movie sets, he has probably never seen a two-dimensional house. Chances are the houses in the pictogram will be interpreted as having three dimensions. If such is the case, the height and width of the 1970 house should be only $\sqrt[3]{4.2}$ equals 1.432 times as great as the same dimensions for the 1946 house, as shown in Figure 5–23.

Figure 5–23 is drawn accurately assuming one is certain that the viewer will definitely conceive the houses as having three rather than two dimensions. But how can we really be certain? As a matter of fact, the 1970 house as shown in Figure 5–23 doesn't seem large enough to me. Could it be that even though I think I am seeing three-dimensional houses, my brain is interpreting the picture as if the houses had only two dimensions? The mere fact that even I, the guy who contrived this example, can be confused on this point, serves to point up a virtually insurmountable problem associated with trying to make charts entertaining—one never quite knows how the viewer will interpret the pictures.

Oh yes. There is also another hazard. The risk always exists that the viewer will glance at the pictures—the houses in the present case—and think, "Fooey! Houses weren't that much bigger in 1970 than they were in 1946!"

Pictures do liven up a chart, but they can so easily convey a false impression that you will be wise to avoid them when constructing charts yourself. When you run across pictograms in magazines or newspapers, you should

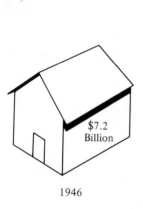
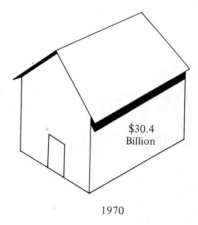

1946   1970

Figure 5–23

examine them with considerable care. Maybe someone is trying to help you get a wrong impression about the thing charted. Even if such is not the case, you can easily mislead yourself by interpreting the pictogram in a way other than that intended by its creator.

Much more could be said about charts, of course. But we have considered some of the most common fallacies. Points made in connection with charts presented here could easily apply to different kinds of charts as well. Let us put this subject to rest and proceed to the subject of descriptive statistical measures. A good place to begin is with one of the most important families of such measures, namely *Averages*.

6

Accommodating Averages

If at first you don't succeed, you're about average.

—ANONYMOUS

The story is told about a traveling man who registered at a luxury hotel. A bellhop carried the man's luggage to his room and waited for a tip. Catching the hint, the man asked, "How large a tip do customers of this hotel usually give you?" "Two dollars is about average," the boy replied without hesitation. The man, obviously surprised, removed two one-dollar bills from his wallet and handed them grudgingly to the bellhop. The boy, pure delight showing on his face, thanked the man profusely and said, "You're the first one ever to come up to my average."

Clearly, the bellhop had his own highly individualistic notion of what an average is. That's the way it is with averages. The word itself originated as a clearly defined insurance term based on the fact that many kinds of data tend to cluster around some central or "representative" value. But today the term has so many diverse meanings that it is hopelessly confusing unless accompanied by supporting credentials of some kind. But confusing or not, the word "average" is still much used by statistician and layman alike. Understanding its many meanings is therefore essential to the budding statistical critic.

Let us dispose of one misuse of the word right now: To speak of an

"average man" or of "average weather," etc., is to speak nonsense even though many people do it. A man may be average with respect to height, weight, income, I.Q., excitability, or any other specific, measurable quality. But the term "average man" conveys nothing except maybe the vague notion that the man in question is not outstanding in any way that meets the eye. Similarly, the weather might be average with respect to temperature, rainfall, or humidity, but not just "average." The term, to have any precise meaning, must be used in conjunction with some specific quantifiable property.

Unfortunately, even when "average" is used to describe some specific quantifiable property, its meaning can still be pretty vague and the figure quite misleading. The reason is that there are many kinds of averages. From a formal statistical standpoint some are better suited to certain situations than others. By the same token, some are better suited to presenting a distorted picture of the data than others.

The three most frequently used averages are the *arithmetic mean*, the *median*, and the *mode*. Less frequently used, but still worth knowing something about, are the *weighted mean*, the *geometric mean*, and the *harmonic mean*. One should obviously not barge into the calculation of an average haphazardly. Instead, he should select the average that provides the least distorted one-figure summary of the data.

The necessity of selecting an appropriate type of average before beginning calculations may seem self evident. Nevertheless, it is a consideration ignored by many people who simply assume that an average is an average and "why make a big thing out of it?" The other side of the story is that there are many people who are keenly aware of the several averages in the statistician's tool kit but are devoid of the professional statisticians's high regard for the truth. This is the group that intentionally selects the most misleading possible average to present the data in a distorted way. We must be wary of the statistical products of both groups—the uninformed and the well informed but unscrupulous—because different averages calculated from exactly the same set of basic data may differ significantly from one another. Moreover, one or more such averages might be perfectly well suited to the data whereas others are definitely inappropriate.

Using the Wrong Kind of Average

Let us assume that the latest issue of *Skylarker* magazine contains an eye-filling, full-page advertisement dedicated to enticing consumer-goods businesses to advertise in the magazine. Assume further that the *Skylarker* ad reads: "Last year, subscribers spent an average of $400 on gifts." The $400 figure may sound impressive, but how much information does it actually convey? The answer is practically none.

I must ask you to accept for the moment the preposterous assumption that the magazine has only seven subscribers: Adams, Baker, Clark, Davis, Elsea, Ford, and Gotlotz. Ignoring Ford and Gotlotz for now, let us suppose that the other five subscribers spent the following amounts on gifts last year: Adams, $48; Baker, $48; Clark, $50; Davis, $60; and Elsea, $70.

When one speaks of "average" without qualifying the term, he is usually assumed to be referring to *the arithmetic mean*—the sum of the figures divided by the number of figures being averaged. In this case, the arithmetic mean is $(48 + 48 + 50 + 60 + 70)/5 = \$55.20$. However, as previously mentioned, several other kinds of averages can be obtained from these same figures, two frequently used alternatives being the median and mode.

The median is simply the figure such that exactly half of the numbers are lower in value and exactly half are higher. Applying this definition to our five numbers, we find that the median amount spent for gifts is $50.00 because 50 exceeds the two 48's but is exceeded by the 60 and 70.

Finally, *the mode* is the figure that occurs most often, namely the $48.00 in the present case. Therefore, we have: mean = $55.20, median = $50.00, and mode = $48.00. But which of these is the most appropriate representative figure? The answer depends entirely upon how the figures are to be used. In this example, it probably matters little which average is used because the three figures are very nearly the same.

But now let's get Ford, a rather affluent gentleman, into the act. Suppose

Ford spent $300 on gifts last year. We now have six figures—48, 48, 50, 60, 70, and 300—to be averaged.

With the introduction of Ford the arithmetic mean becomes a fairly impressive $96.00, exactly double the mode (still only $48.00) and considerably higher than the median, $55.00 (the two competing figures, 50 and 60, having been summed and divided by 2 as is traditional when working with an even number of figures). It now matters a good deal which average is used. The arithmetic mean of $96.00 is higher than five figures and lower than only one, a condition suggesting that its use as a representative measure could be quite misleading.

Pushing the example to an even more outlandish extreme, let us now bring in Gotlotz who spend $2,224.00 on gifts last year. We find that the arithmetic mean jumps to $400.00, the figure appearing in our hypothetical advertisement, whereas the median is a mere $60.00 and the mode is still only $48.00, the same as always. Whether the median or the mode would be preferable as a representative figure is a matter for debate. One thing is crystal clear, however: The arithmetic mean is too greatly influenced by the amount spent by Gotlotz to be representative of all.

We can now relax the assumption about the magazine's having only seven subscribers without negating the point just made. Whether the number of subscribers is few or many, the related gift-giving figures will not necessarily be distributed symmetrically as in the following histogram[1] where mean, median, and mode are equal:

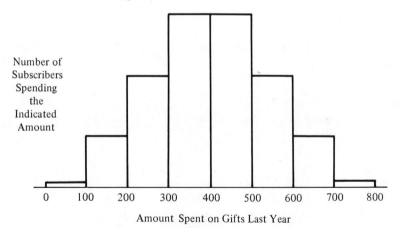

Number of
Subscribers
Spending
the
Indicated
Amount

Amount Spent on Gifts Last Year

Figure 6–1

[1] A histogram is merely a specific kind of bar chart. To construct a histogram one organizes the basic data into classes and shows, by means of bars, how many of the numbers are found in each of the various classes. Most good books on statistical techniques show how to construct and interpret histograms. Such knowledge, while desirable, is not essential to an understanding of the points made here where histograms are used for purposes of illustration.

This symmetrical distribution is probably the kind the copywriter of the ad hoped we would envision. But because expenditures for gifts are tied to individual income levels and because incomes are distributed in a highly skew (or asymmetrical) manner, the true distribution is probably more like the following one:

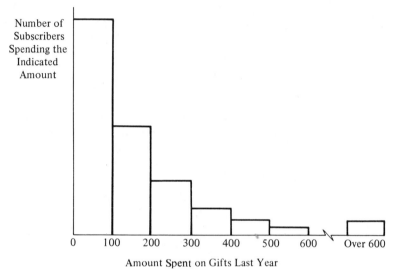

Figure 6–2

If so, the arithmetic mean would be misleading as a representative figure for reasons already indicated.

Let us switch examples for a moment and consider another situation where the arithmetic mean can be misleading in quite a different way. The arithmetic mean is a measure of central tendency; where no central tendency exists its use is of questionable worth. The following graph, for example, is roughly similar to distributions of percent cloudiness for many American cities:

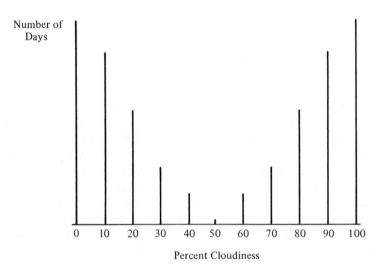

Figure 6–3

This chart shows that 0-percent and 100-percent cloudiness both occur more often than 10- or 90-percent cloudiness, and the latter, in turn, occur more often than 20- or 80-percent cloudiness, and so forth. The arithmetic mean of 50 percent (assuming a perfectly symmetrical distribution) could give the false impression that, typically, the sky is about half covered with clouds when in fact it is more likely to be either completely covered with them or completely free of them. This distribution could best be described as bi-modal with modes at zero and 100.

If all this mean, median, and mode business has you a little confused, don't feel bad for you're in excellent company. These various averages are confusing to quite a number of highly sophisticated people. Take, for example, the following naive statement made by an eminent scholar and scientist, Dr. Karl A. Menninger:

> Fortunately for the world, and thanks to the statisticians (for this, of course, is a mathematically inevitable conclusion), there are as many people whose intelligence is above the average as there are persons whose intelligence is below the average. Calamity howlers, ignorant of arithmetic, are heard from time to time proclaiming that two-thirds of the people have less than

average intelligence, failing to recognize the self-contradiction of the statement.[2]

Although there might—I repeat, might—indeed be as many people with above-average intelligence as there are with below-average intelligence, the self-contradiction to which Dr. Menninger refers simply doesn't exist—not if he has in mind the arithmetic mean when he speaks of "average." His remark would be generally correct only if he were speaking of the median.

If Dr. Menninger were to apply his faulty reasoning to our gift-spending figures, his conclusion would almost certainly be dead wrong because, as I have argued, the data are probably distributed in a skew manner with the arithmetic mean being substantially higher than the median and mode. In this case, far more people would be spending an amount below the average than would be spending an amount above the average and the arithmetic mean would present a misleading summary of the basic data.[3] When the data are distributed more or less symmetrically, the mean and median (and perhaps the mode, though not necessarily) will be sufficiently close that it matters little which is used. The point for you to remember as a consumer of statistics is: When you are told neither what kind of average was used nor the manner in which the data are distributed, you will be wise to view the alleged average with a jaundiced eye; someone just might be using the arithmetic mean, intentionally or unintentionally, as an instrument for exaggerating a point or conveying a distorted impression.

Other Kinds of Averages

There are other kinds of averages in addition to the mean, median, and mode, three of which I feel obligated to touch on. These are the weighted mean, the geometric mean, and the harmonic mean.

The Weighted Mean

Suppose you are told that 20 students in a certain physical education class ran the hundred-yard dash in times (expressed, for simplicity, to the nearest whole second) ranging between 11 and 20 seconds. Only one student ran the distance in 11 seconds; two in 12 seconds; two in 13 seconds; ten in 15 seconds; and five in 20 seconds. What was their average speed? I hope you

---

[2] *The Human Mind*, Third Edition (New York: Alfred A. Knopf, 1945), p. 199.

[3] I should probably mention that statisticians do many interesting things with the arithmetic mean when analyzing data from samples for purposes of statistical inference even though such data might be a far cry from symmetrical. This fact, however, does not negate the points made above regarding the probable impropriety of using the arithmetic mean as a purely descriptive measure of a set of skew data.

resisted the temptation to say 14.2 seconds: $(11 + 12 + 13 + 15 + 20)/5$ $= 14.2$. The fallacy in taking this average lies in the fact that the number of students running the distance in the five specified times is not the same. The correct process of averaging would be

$$\frac{(1)(11) + (2)(12) + (2)(13) + (10)(15) + (5)(20)}{1 + 2 + 2 + 10 + 5} = \frac{311}{20} = 15.55.$$

This procedure yields the same results that would be obtained if each student's time was listed separately (that is, 11, 12, 12, 13, 13, and so forth), summed and divided by 20, the number of students.

The Geometric Mean

The geometric mean is used to average rates of growth where a measurement is dependent upon previous measurements, the growth building upon itself—that is, compounding. To illustrate: Suppose the population of a certain city had been two million in 1950 and had grown to four million by 1970. What would be your best guess of the population figure for the year 1960? Three million (2 plus 4 divided by 2) would almost certainly be too high because the increase in population had been more rapid in the later years of the period than it would have been in the earlier years. A more nearly accurate guess would be 2.8 million, obtained by multiplying 2 by 4 and taking the square root of the resulting product of 8.[4] When a series of numbers grows in a compound manner, the geometric mean will generally be lower than the arithmetic mean and more appropriate.

The Harmonic Mean

At some time or other you have probably seen a variant of the following puzzle and, like many people, have been tempted to use the arithmetic mean to solve it: A motorcyclist wants to ride to the top of a hill and back at an overall average speed of 60 miles per hour. If his mean speed going up is 30 miles per hour, what must his mean speed be coming down? Ninety miles per hour? I'm afraid not. There is actually no speed for the second half of the trip that can be great enough to result in a 60 miles-per-hour trip over-all. If the total distance is d miles, an overall speed of 60 m.p.h. requires that the trip be completed in d minutes. But traveling half the distance at 30 m.p.h.

---

[4] If three interdependent figures are to be averaged, they are multiplied together and the cube root of the resulting product is obtained; if four, the four figures are multiplied and the fourth root of the resulting product is obtained; and so it goes. In general,

$$\text{G.M.} = \sqrt[n]{(X_1)(X_2)\ldots(X_n)}.$$

means that he has already used up the entire d minutes. The harmonic mean is the tool by which problems of this kind are solved.[5]

Although the various averages discussed above serve as useful one-figure summaries of a set of data, when used alone they can also conceal important information. An average, for example, tells nothing about the amount of dispersion in the data, a shortcoming whose consequences will be explored in the next chapter.

[5] Since you are probably no more interested in the harmonic mean than I am, let's do the dirty work down here in a footnote which can be easily skipped without an ensuing sense of guilt.

The harmonic mean is defined as

$$\text{H.M.} = \frac{n}{1/X_1 + 1/X_2 + \cdots + 1/X_n}.$$

For the problem under discussion:

$$\text{H.M.} = \frac{2}{1/30 + 1/X},$$

where 2 represents the two times that one-half the distance must be traveled, $\frac{1}{30}$ is the reciprocal of the known going-up speed, and $1/X$ is the reciprocal of the unknown coming-down speed. If H.M. is to be 60, its reciprocal must be $\frac{1}{60}$, which means that the sum of the reciprocals of the two speeds must be $\frac{1}{30}$. Since the reciprocal of the first speed is $\frac{1}{30}$ the reciprocal of the second speed must be zero, an impossibility since no finite number has such a reciprocal.

7

Ignoring
Dispersion

*When I was a young man practicing at the bar, I lost a great
many cases I should have won. As I got along, I won a great
many cases I ought to have lost; so on the whole justice was
done.*

—LORD JUSTICE MATTHEWS

Two statisticians were drafted into the army and soon found themselves
fighting side by side on the front lines. Simultaneously spotting an enemy
soldier, both statisticians raised their rifles and fired. One statistician fired a
foot too far to the right, the other a foot too far to the left. Turning toward
each other, pride of accomplishment written all over their faces, they shook
hands vigorously and said, "Congratulations!"

Needless to say, little was gained from knowing that, on the average, the
enemy soldier was dead. In a case like this it is the trivial details that matter
most, such as the fact that the enemy was really very much alive and preparing
to return fire while the statisticians were celebrating their imaginary achievement.

The point of this far-fetched story is that sometimes dispersion is more
important than an average. By dispersion I mean the amount of scatter in
the data—that is, the extent to which the individual numbers differ from one
another. For example, the set of numbers 3, 3, 3, 3, 3 has no dispersion. The
set 1, 3, 5, 7, 9 does have dispersion but not as much as the set 1, 5, 10, 15, 30.
Many formal measures have been developed to express dispersion as a number. These include the range, mean deviation, standard deviation, variance,

interquartile range, quartile deviation, coefficient of variation, and so forth. For our purposes, there is no need to go into detail about these measures.[1] We must, however, consider some of the consequences of ignoring dispersion altogether.

Why Dispersion Matters

Why need we concern ourselves with dispersion? Because the individual observations may differ substantially from one another and the difference might make a difference.

Suppose, for example, that you know someone faced with the terrifying prospect of taking a course in statistics sometime soon. Let us say he comes to you for advice about which of two professors—let's call them Professor Frank and Professor Stein—he should choose. After some questioning you learn that his state of mind is such that he will be overjoyed just to get out of the course with a C. Also assume, outlandish though it undoubtedly is, that the grade is the only thing about the course that matters to him. Which professor would you advise him to choose? Actually, this is an unfair question at this point since I have told you nothing about Professors Frank and Stein. So let's set it aside until you know more about the two men dishing out the grades.

A little research reveals that the average grade given by each of these professors is C. Would it then be a matter of indifference which professor your acquaintance studied under? Not necessarily. Suppose that Professor Frank had seldom given a grade other than C and that Professor Stein had seldom given a grade other than A or F and these he has given in about equal numbers. Professor Frank, who apparently judges the great majority of students to be just so-so, would be the logical choice. Granted, getting a grade above C would be difficult, but so would getting a grade below C. Professor Stein, who seemingly views students as either excellent or poor, should be avoided even though a grade of A would be a distinct possibility; a grade of F would also be a distinct possibility. To repeat the principal point of this section: Sometimes an average by itself just doesn't provide enough information to help with rational decision making.

Examples of faulty conclusions resulting from ignoring dispersion are fairly plentiful. One oft-cited story has to do with an event during a civil war between two warlords in China in the 1920's. Upon their arrival at the bend in a river one of the warlords and his troops discovered that there were no boats available for the crossing. The warlord, remembering that a geography book he had once read stated that the average depth of the water at that

---

[1] However, I do suggest that, once you have become aware of the possible hazards of ignoring dispersion, you invest some time in acquainting yourself with the details of such measures. All respectable statistics textbooks treat them at length.

time of year was less than two feet, gave the order to cross on foot. After the crossing the warlord discovered, to his astonishment, that several hundred of his soldiers had been drowned. Although the river did indeed average less than two feet in depth, in some places it was considerably deeper than that and, as it happened, over the heads of many shorter soldiers.

An unconfirmed story concerns a quartermaster in the army who reported that the average food consumption per soldier in the field was adequate. He had some explaining to do when several men turned up with symptoms of malnutrition because they had received amounts considerably below the average while others had stuffed themselves.

A former student of mine attended a meeting where the feasibility of establishing a new bank in a nearby suburb was discussed. The main spokesman for those in favor pointed out that the average increase in population in that suburb over the most recent ten years was about 500 people—plenty, he asserted to justify another bank. What he failed to mention was that ten years ago the immigration of people into this suburb was about 3,500 a year and that the amount of immigration had been declining steadily to a mere 50 people, give or take a few, during the most recent year. The average over the ten-year period might well have been about 500 people, just as this proponent claimed, but a single year accounted for 70 percent of the total—and that year was the one furthest removed from the time of the discussion.

Changing the subject a little, sometimes an indication of dispersion is omitted from charts designed to show the relationship between two variables. Frequently, researchers will attempt to express the relationship between two or more variables by means of a mathematical function, a perfectly legitimate statistical procedure when mixed with a generous helping of common sense.

But the relationship shouldn't be presented this way

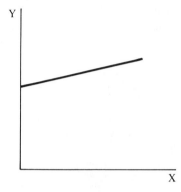

Figure 7–1

a situation suggesting a perfect relationship, when it actually looks like this

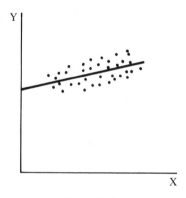

Figure 7–2

More will be said about statistical relationships in Chapter 13. For now, simply remember to be wary of any line purporting to describe a statistical relationship when no scatter (no messy little dots) is shown around the line. The absence of dots implies a perfect relationship, something that doesn't exist outside of mathematics textbooks.

Comparing a Single Observation with an Average

Sometimes a single observation is compared with an average with rather dramatic results—dramatic, that is, as long as no mention is made of dispersion. For example, an acquaintance might boast to you that he can do six more pushups than the average of all other students in his physical education class. What has he really told you? Only that he is above the class average in this respect. He certainly hasn't said that he is the best in the class, or even one of the best, but his boast could easily be interpreted in that way. Without some knowledge of the amount of dispersion around the average, you can't even make a very shrewd guess as to where he stands in relation to the others. For all you know, practically half of the class[2] might be able to do several more pushups than he.

A cigarette company advertised that one of its brands had been found by a U.S. Government report to be 43 percent lower in tar than the average of all other menthol cigarettes. I trust you can now criticize this claim without any help from me.

Always be wary of comparisons where a single observation is compared with an average unless you know for a fact that there is precious little dispersion in the data. In the absence of knowledge about the amount of dispersion, such comparisons simply don't tell you much of anything.

Sometimes percents don't tell you much of anything either. How's that for a smooth transition into the next chapter?

---

[2] Theoretically, even more than half of the students might be able to surpass your acquaintance. If the modal number of push-ups were very high and the arithmetic mean were pulled down by a few students who could do practically none, your acquaintance might be able to do six more push-ups than the average and still be below the majority of class members.

8

Puffing Up a Point with Percents

He uses statistics as a drunken man uses lampposts—for support rather than illumination.

—ANDREW LANG

Probably no family of statistical measures has been used more often or more effectively or in a wider variety of ways over the years to prop up wobbly arguments than that of the percent. Practically any situation can be quickly improved or worsened—on paper—just as long as there are numbers around that can be converted into percents.

It is probably also safe to say that few statistical measures surpass the percent in the frequency with which honest mistakes are made. The existence of both honest misuses and wanton abuses of percents seems all the more regrettable when we realize that a percent—any percent—appears magnificently respectable. The percent sign itself suggests a kind of finality, as if to say, "Here. The problem has already been worked out for you." If a percent figure has a decimal in it, so much the better from the misuser's standpoint, for then only the most hardened cynic would stoop to cast aspersions on so noble a measure—or so many people seem to think (and with some justification). In this chapter we survey a considerable variety of ways that percents get misused.

Some Messy Distinctions

One reason why percents get misused is that much confusion exists over the terms "percent," "percent change," and "percent points of change." Distinguishing among these terms can be facilitated by an example: Suppose that on July 1, 1965, you bought 100 shares of a common stock— let's call it South Sea Bubbles—for $20.00 a share. Suppose also that you bought this particular stock for its growth potential. In fact, so sure have you been of its growth possibilities that you have checked the stock's market price only once a year since the day you bought it; this you have done on July 1, or the nearest trading date, of each year from 1965 through 1972. (Please humor me on this. I'm just trying to keep the example simple.) Finally, suppose that the stock's market prices have been as shown in Table 4. Now let's see what kind of sense can be made out of the sometimes tricky terminology of percents.

Table 4. Market Price per Share of South Sea
Bubbles on July 1 of Each Year

Year	Price
1965	$20
1966	30
1967	35
1968	40
1969	30
1970	20
1971	30
1972	35

Source: Hypothetical data.

Percent

A convenient way of keeping track of how favorably or unfavorably the July 1 market prices compare with your purchase price would be to convert the figures in Table 4 into so-called simple relative index numbers. This is done by expressing the $20.00, $30.00, $35.00, etc., as percents of the 1965 base figure of $20.00—that is, by dividing each number in the list by $20.00 and multiplying the resulting quotient by 100.[1] The percent figure for 1966,

---

[1] The general expression for finding the percent figure in a situation like this would be

$$\text{Percent} = \frac{X_L}{X_E} \cdot 100,$$

where X_E is the value in the earlier period and X_L is the value in the later period.

for example, would be obtained by dividing the market price of $30.00 by the base value of $20.00 and multiplying the resulting 1.5 by 100, giving you 150 percent. If you were to follow this same procedure for each market price in the list, the results would be as shown in the right-hand column of Table 5.

Table 5. Market Price and Percent Relative of a
Share of South Sea Bubbles on July 1
of Each Year

Year	Price	Percent Relative (1965 = 100)
1965	$20	100
1966	30	150
1967	35	175
1968	40	200
1969	30	150
1970	20	100
1971	30	150
1972	35	175

Source: Hypothetical data.

Figures in the right-hand column of Table 5 tell you at a glance that the market price in mid-1966 was 150 percent of the market price in mid-1965; the market price in mid-1967 was 175 percent of that of mid-1965; and so forth.

Percent Change

The percent change, like the simple percent relative, is a ratio with a base-period denominator. Unlike the percent relative, however, the numerator is the *change* between two points in time.[2] Hence, the percent change in price between mid-1965 and mid-1966 was

$$\frac{30 - 20}{20} \cdot 100 = 50 \text{ percent};$$

the percent change from mid-1965 to mid-1967 was

$$\frac{35 - 20}{20} \cdot 100 = 75 \text{ percent; and so on.}$$

[2] In symbols:

$$\text{Percent change} = \frac{X_L - X_E}{X_E} \cdot 100,$$

where X_E represents the value in the earlier period and X_L the value in the later period.

We are now in a position to note two peculiarities of the percent change. First, percent changes are not reversible. Notice, for example, that between 1965 and 1968 the market price of the stock doubled, climbing from $20.00 to $40.00. Then between 1968 and 1970 it dropped right back to the $20.00 you paid for it (an event casting doubt on its merits as a growth stock). The first increase amounted to 100 percent — $[(40 - 20)/20] \cdot 100 = 100$ percent —, but the decline back to $20.00 was only 50 percent — $[(20 - 40)/40] \cdot 100 =$ − 50 percent —, not 100 percent as one might naturally assume. The reason, of course, is that although the absolute magnitude of the change remained constant, the base value doubled. Remember this pecularity of the percent change the next time you read in the newspaper that because of White House pressure an announced five-percent price hike in some steel products was rolled back. Don't let a journalist convince you that the rollback amounted to a five-percent drop from the new base.

A second peculiarity of the percent change is that decreases greater than 100 percent are not possible in situations where the original figures are all positive. Often statements are made to the effect that a decrease was greater than 100 percent when any such calamatous drop is clearly impossible. In a *Newsweek* article, for example, the statement was made that Mao-Tse Tung had slashed salaries of some Chinese government officials by 300 percent. An alert reader inquired, "How much of the original salary is left to be further slashed after it has been reduced by 100 percent?"[3] The editors admitted that the correct figure was 66.67 percent, not 300.

Percent Points of Change

Returning to our common-stock figures, we note that the percent relative for 1966 was 150 and for 1967 it was 175. The percent points of change therefore was 25, a figure obtained simply by subtracting the smaller percent figure from the larger.[4] It would be wrong to say that the percent increase was 25. The percent increase was actually

$$\frac{35 - 30}{30} \cdot 100 = 16.7.$$

As you might suppose, a figure representing percent points of change is often interpreted, incorrectly, as if it were a percent change. To cite an example close to home: A colleague of mine, recently testified before a public utilities commission whose task it was to evaluate a request for a rate increase.

---

[3] January 16, 1967, p. 6.

[4] This simple procedure can be expressed as

$$\text{percent}_L - \text{percent}_E.$$

Included in one of the arguments voiced by a spokesman for the company was the figure of 4.8 percent said to represent the most recent monthly reading on the rate of inflation for the entire country. This figure was obtained, the company spokesman explained, by taking the change in the Consumer Price Index since the preceding month and multiplying that change by 12 to put the figure on an annual-rate basis. Since the change in the Consumer Price Index had been 0.4, 12 times that figure came to 4.8. This last figure was referred to as the annualized percent change. Multiplication by 12 was justifiable enough, but use of the 0.4 was not. Because the Consumer Price Index had been substantially above 100, a change of 0.4 percent points was not the same as a percent change of 0.4, as my colleague explained to the commission. The true percent change was closer to 0.3 which, annualized, came to only 3.6 percent.

Carelessness in Computing Percents

In my stock-price example, I treated the various percent measures as if they were always calculated over time. Such is not the case. Any two numbers can be compared with respect to relative size through use of a percent of some kind. For example, suppose that you want to compare the official weights of two heavyweight boxers, A and B. A weighs in at 210 pounds and B at 225 pounds. A's weight as a percent of B's would be computed as follows:

$$\frac{\text{A's weight}}{\text{B's weight}} \cdot 100 = \frac{210}{225} \cdot 100 = 93.3 \text{ percent.}$$

If you want to know by what percent A's weight is less than B's, you can simply subtract 100 from the 93.3, giving -6.7 percent. An alternative way of getting the latter figure is to proceed in a manner analogous to the calculation of a percent change. Begin by determining the difference between the two weights and use that difference as the numerator. The denominator is B's weight. Therefore,

$$\frac{\text{A's weight less B's weight}}{\text{B's weight}} \cdot 100 = \frac{210 - 225}{225} \cdot 100 = -6.7 \text{ percent.}$$

Alternatively, B's weight as a percent of A's would be

$$\frac{\text{B's weight}}{\text{A's weight}} \cdot 100 = \frac{225}{210} = 107.1 \text{ percent.}$$

B would be heavier than A by

$$\frac{\text{B's weight less A's weight}}{\text{A's weight}} \cdot 100 = \frac{225 - 210}{210} \cdot 100 = 7.1 \text{ percent.}$$

Simple though such percent calculations are, careless errors are made all too often. For example, an advertisement for a famous brand of paper towel stated that this brand absorbs 50 percent more than competitive brands because it is two layers thick rather than just one. The truth of the claim aside, the correct percent is 100, not 50:

$$\frac{\text{Number of layers for this specific brand less number of layers for other brands}}{\text{Number of layers for other brands}} \cdot 100$$

$$= \frac{2-1}{1} \cdot 100 = 100 \text{ percent.}$$

The writer of a letter to *The New York Times* commenting on an article about who has it best, men or women, said that he trusted statisticians will find that women give birth to 100 percent more babies than men. Women do, of course, give birth to 100 percent of the babies born, but their lead over men is vastly greater than a mere 100 percent.

Confusing Percent of Total with Numerical Level

Between 1950 and 1951 spending by state and local governments slipped from about seven percent of the nation's gross national product to around 6.5 percent. A natural tendency is to ask what happened to weaken spending by governments on the state and local levels. The answer, in this specific instance, is nothing weakened it, except in a relative sense. State and local government spending actually increased from $19.5 billion in 1950 to $21.5 billion in 1951. However, the outset of the Korean War led to substantial jumps in spending by the federal government, businesses, and consumers. State and local governments increased their spending by too little to maintain their previous share of the gross national product.

Along this same line, I recently told my business forecasting class that property taxes had been declining in relative importance as a source of funds for state and local governments. Some students interpreted this remark to mean that property tax collections had been declining absolutely. But such was not the case; such collections had been increasing but at a slower rate than personal taxes, sales taxes, and federal grants-in-aid. Hence, a decline in the relative importance of property taxes had occurred.

Comparing Percents Based on Levels
Rather than Percents Based on
Changes in the Level

In Chapter 1, I quoted some excerpts from Eugene Lyons' *Workers' Paradise Lost* wherein he reveals some statistical sleight of hand practiced by the Kremlin in an effort to make Stalin's first Five Year Plan look like a success rather than the failure it really was. One example had to do with steel output. Steel output in 1928 was 4.2 million tons. The Plan foresaw an increase to 10.3 million tons. Actual production in the final year was 5.9 million tons, which represented an increase of 1.7 rather than the planned 6.1 million tons.

The proper way of comparing the output obtained with the output planned is as follows:

$$\frac{\text{Change obtained in steel output}}{\text{Change planned for steel output}} \cdot 100 = \frac{1.7}{6.1} \cdot 100 = 27.9 \text{ percent.}$$

The Kremlin, however, compared *level* obtained with *level* planned. Hence,

$$\frac{\text{Level of steel output obtained}}{\text{Level of steel output planned}} \cdot 100 = \frac{5.9}{10.3} \cdot 100 = 57.3 \text{ percent.}$$

The trouble with such an approach is that, if steel output hadn't increased at all, the Plan would have been fulfilled to the tune of $(4.2/10.3) \cdot 100 = 40.8$ percent. Even if output had declined by 50 percent from the 1928 base, the Plan would have been fulfilled, according to this approach, to the extent of $(2.1/10.3) \cdot 100 = 20.4$ percent. Such a situation would amount to progress by going backward.

Economic forecasters often commit this fallacy in an effort to make their track records look better than they really are. Suppose, for example, that the nation's gross national product in a given year is $1,000 billion and the forecaster foresees an increase to $1,100 billion for the year coming up. Suppose further that the forecast year actually registers an increase to $1,050 billion. The honest thing for the forecaster to do would be to claim 50 percent accuracy in his prediction. That is, he expected an *increase* of $100 billion, but the actual *increase* was only $50 billion. However, he might be more strongly disposed to claim 95 percent accuracy. He foresaw a *level* of $1,100 billion and the actual *level* was $1,050 billion—hence, 95 percent accuracy.

Adding and Subtracting Percents

The following tall tale, dreamed up by a prankster named C. Louis Mortinson, was published as fact in the February 9, 1935 issue of *Newsweek*:

Logic: Lester Green of Prospect, Conn., puts two setting hens on his automobile motor cold nights. "A setting hen's temperature is 102," Green explained, "and consequently two hens is 204. With this heat the engine is sure to start the first time it kicks over."

The stationery of the Omak, Washington, *Chronicle* furnishes this information:

Area code	509
Telephone	826–1110
P.O. Box	553
Zip Code	98841
State sales tax no.	C243292
Federal tax no.	91–0664–86
Street address N. Main	109
Bank account no.	26128108
Business established in	1910
Our total number	25,800,918[5]

Although the preceding examples have nothing to do with percents, they do point up, so that the absurdity is transparent, the fact that some numbers are simply not additive. Oh sure, any numbers can be added together, but under some circumstances the totals are at best worthless and at worst woefully misleading. This point applies with a vengeance to percents. Percents frequently are added, subtracted, averaged, and what have you, without justification and with an uncomfortably high probability that the underlying fallacy will go undetected.

Suppose, for example, that Tom, Dick, and Harry are the president, vice-president, and secretary-treasurer, respectively, of a company whose board of directors has just voted the top officers a salary increase. Tom, let us say, has been voted a five-percent increase, Dick a ten-percent increase, and Harry a 15-percent increase. Is the total salary increase received by these three officers 30 percent, the total of the three percent increases? No. Nor is it 10 percent, the average of the three. To determine the true combined increase, we must know the base values. Let us assume that before the raises the salaries were

Tom	$100,000
Dick	20,000
Harry	10,000
Total	$130,000,

which means that after the raises go into effect the salaries will be

Tom	$105,000
Dick	22,000
Harry	11,500
Total	$138,500.

[5] Presented here by permission of the Omak, Washington, *Chronicle*.

The combined percent increase, therefore, is

$$\frac{138,500 - 130,000}{130,000} \cdot 100 = 6.5,$$

a figure which is, in effect, a weighted mean of the three percent increases.

Picture yourself as the president of a manufacturing concern. You have received a report from the accounting department showing that production costs have increased 12 percent during the past two years. You know that the selling price of the product has increased by only eight percent over the same period. Should you conclude that profits have decreased by four percent? You could be quite wrong if you do. Assume, for example, that the cost of production during the base period had been $1.00 a unit and that the selling price had been $2.00 a unit. The 12-percent increase on the $1.00 base figure brings the cost of production to $1.12 while the eight-percent increase on the $2.00 figure brings the selling price to $2.16. The profit figure, consequently, increases from $1.00 to $1.04 a unit, a four-percent increase rather than the presumed four-percent decrease.

A good rule is never add, subtract, or average percents unless you are absolutely certain that all percents relate to the same base figure.

Fallacies Related to Selection of a Base Value

Some of the most colorful examples of misuses of percents arise because the choice of a base is in some way improper. Perhaps the base figure is extremely small. Or maybe the base figure and/or the base period are handpicked with a view to encouraging an incorrect interpretation.

Percents Misleading Because of Small Base Values

The classical example of a misleading percent resulting from too small a base figure is the one about the marrying habits of women students at Johns Hopkins University. Some decades ago Johns Hopkins broke a precedent and began admitting female students. Shortly thereafter it was reported that 33 1/3 percent of the women students who enrolled had married members of the faculty. The story lost some of its punch when someone disclosed that only three women had enrolled in that first year and one had married a faculty member.

A former student of mine related an interesting story about the way two local newspapers reported crime statistics for a small city near his home. According to the morning newspaper, homicides in that particular city had increased by 60 percent over the previous year! Sixty percent! It sounds as

if the residents were being slaughtered en masse. The student telling the story said he had visions of the place becoming a ghost town before the next batch of crime statistics was released. Fortunately, the afternoon newspaper set matters straight by reporting the actual number of homicides. It stated that homicides had increased from five to eight in a year's time. Eight, of course, does represent a 60-percent increase over five; therefore, the morning paper wasn't actually lying. Still, the 60-percent figure was misleading appearing as it did unaccompanied by the actual number of homicides contributing to the large percent change.

When a base figure is small, as it was in this example, a small numerical increase can be expressed as a very large, and probably most impressive, percent change. Consider, for example, a numerical change of 3 between two points in time. Is 3 a large or a small change? The answer depends entirely on the size of the base figure. If the base figure had been, say, 1000, the percent change would be a mere 0.3. If the base figure had been 1, on the other hand, the percent change would be a startling 300. Everything hinges on the size of the base figure, and, if it is extremely small, grotesque results can occur. Beware of percent changes, or any percents, for that matter, when the base values aren't given. If the supplier of a percent has nothing to hide, he will gladly reveal the actual figures on which the calculation was based.

Here is a variation on the use of small base values: Early in 1968 shortly before the ten-percent surtax went into effect, a prominent political figure warned that an increase in taxes accompanied by a sharp cut in federal govern-ment spending might result in a 30-percent boost in the unemployment rate. Now 30 percent sounds intolerably large when the breadwinning power of fellow humans is at stake. If you will reflect on this figure for a moment, you can easily conjure up mental images of souplines and other symbols of severe depression.

But remember the speaker didn't say that the unemployment rate would be 30 percent; he said that the unemployment rate would increase by 30 percent. What he didn't mention was a fact available to anyone with suf-ficient interest to look it up, namely that the unemployment rate at that time was a mere 3.6 percent. This rate of unemployment amounts to over-full employment when measured against the traditional yardstick of 4.0 percent being equated with full employment. A 30-percent increase would have pushed the unemployment rate to a still respectable 4.7 percent. I don't care to argue whether 4.7 percent is or is not too much unemployment. The point is that the 30-percent figure was apparently used to puff up a point. The politician could have said that the unemployment rate would increase from 3.6 to 4.7 percent. Or he could have said that the unemployment rate would rise by about one percent point. But these are not very colorful ways of putting the point across. He made his point pack about as much wallop as he possibly could by speaking in terms of a 30-percent increase.

Choosing a Convenient Base Figure

Another way in which percentages are sometimes used to put over a point as dramatically as possible is this: If the statistical deceiver wants to make a figure seem small, he can express it as a percent of something large; if he wants to make it seem large, he can express it as a percent of something small. Margaret Knight cites the following example of a situation in which both alternatives were used:

> I once attended a debate on the sterilization of mental defectives in which both sides made use of this device—with the result that each side was able to use the same figure quite effectively in support of its own case. Both the pro-poser and opposer of the motion were apparently agreed to this extent—that there are about 400,000 mental defectives in Britain, and that if a sterilization policy were adopted, their number would be approximately halved in 50 years.
>
> The proposer said: "There are about 400,000 mental defectives in the country—equal to about 25 percent of the whole population of Wales. Sterili-zation would approximately halve the number in 50 years."
>
> The opposer said: "Today mental defectives number less than one percent of the population. Fifty years of sterilization would probably reduce this

proportion to one-half of one-percent—but is it worth it?" Both statements are based on the same "raw" figures, but they certainly give a very different impression.[6]

Similarly convenient base choosing shows up in debates about the advisability of developing stricter gun control laws. Opposers of stricter laws frequently argue for stronger penalties for those who misuse guns and point out that 99 percent (or some other similarly startling figure) of the honest gun owners don't need regulation. Proponents of stricter laws counter with arguments to the effect that 80 percent (or something in that neighborhood) of the murders committed in the United States are performed by normally law abiding citizens. One argument uses the large set of gun owners as a base; the other, the much smaller set of gun murders.

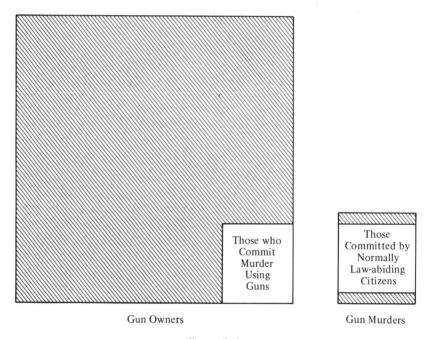

Figure 8–1

Convenient selection of base figures also has a way of popping up just about every time a proposed income tax increase is debated. In 1951, to choose one fairly typical case, The Ways and Means Committee of the House of Representatives considered raising personal income tax rates three percentage points across the board. The tax scale, at that time graduated from 20 percent

[6] "Figures Can Lie," *Science Digest*, September 1951, p. 55. (Condensed from *The Listener*—London)

up to 91 percent, would, after the increase, run from 23 to 94 percent. Some critics attacked the proposal on the grounds that it was aimed at "soaking the poor" because a three-percent point increase on the poor man's 20 percent represented a 15-percent increase while three percent points on the rich man's 91-percent would be a trifling 3.3 percent. Those on the other side of the argument asserted just as vehemently that it was a "soak-the-rich" measure since the poor man's take-home pay would be reduced from 80 to 77 cents on his dollar of income, or only 3.75 percent, while the rich man's take-home pay would be cut from nine to six cents, or 33.33 percent. (The upshot of all this was that the committee compromised by increasing taxes 12.5 percent across the board. This resulted in a minimum rate of 22.5 percent but, unfortunately, boosted the maximum rate to 102.4 percent. It was subsequently cut to 94.5 percent.)[7]

In situations where someone is free to select a base figure to his liking, it is seldom that the one selected is absolutely wrong. But the mere fact that the person doing the selecting can exercise such unrestrained discretion suggests that arguments supported by percent figures should be evaluated on their own merits as much as possible without the percents themselves being allowed to influence our judgment.

Choosing an Unrepresentative Base Period

Unfortunately, it is easier to say what isn't a representative base period for calculating percent changes than to say what is. Here are two examples of what are not representative base periods:

During the Nixon Administration's Second Inflation Alert, reference was made to a 16-percent increase in the average price of regular grade gasoline in the week ending November 17, 1971. A spokesman for the oil industry called the Administration to task for using the 16-percent figure as if it were a typical example of what was happening to prices in this major U.S. industry. The spokesman's argument is quite impressive and, apparently, valid:

> . . . The weekly price change cited was particularly inappropriate . . . because it compared the November 17 price of 25.66 cents per gallon to the November 10 price of 22.14 cents, which happened to be the lowest weekly price figure published by the Journal [*The Oil and Gas Journal*] since February 14, 1969, a period of 21 months.
> The misleading nature of such week-to-week comparisons . . . is further illustrated by the fact that had the Second Inflation Alert been issued a week earlier you would have been comparing the 22.14 cent price of November 10

---

[7] Cited in William A. Spurr and Charles P. Bonini, *Statistical Analysis for Business Decisions* (Homewood, Illinois: Richard D. Irwin, Inc., 1967), pp. 14–15.

with the 23.35 cent average of November 3, and the result would have been a 5.2 percent drop in the gasoline price.

The average price of gasoline for the last three years has fluctuated in the order of 23 cents to 25 cents. A one-cent increase in this price would be about 4 percent, not 16 percent.[8]

Earlier in this chapter, I described an experience of one of my colleagues who testified before a public utilities commission assigned the task of evaluating a request for a rate increase. In that example, I told how he had caught the company spokesmen in the mistake of referring to a percent point increase as if it were a percent increase. Clearly, this error could have been merely an honest mistake. More dramatic, and more likely to have been an intentional abuse of percents, was an argument utilizing average annual percent changes in the company's earnings. Here is what took place: The proponents produced figures showing that their company's average annual percent increase in earnings since 1961 was significantly below similar earnings growth rates for a broad cross section of other companies. If my colleague had conceded that 1961 was an appropriate base year, the argument would have stood up pretty well. However, this particular base year was, from all appearances, one that had been carefully handpicked to make the company's earnings growth seem particularly sluggish relative to that of the other companies.

In 1960 the public utility had been granted a rate increase which showed up in the form of a substantial jump in earnings in 1961. When average annual percent rate of growth was computed using this inflated earnings figure as a base, the comparison used by the proponents did indeed show their company's earnings growth lagging behind that of the sample of other companies. My colleague, however, did some figuring of his own. He computed average annual percent rates of growth in earnings for this public utility and for the same sample of companies that the proponents had used; but he used 1958 as a base year, and then 1959, and then 1960, and so forth, up through 1967. He found that using any other year than 1961 as a base, the average percent increase in this public utility's earnings surpassed the average for the sample of other companies. The public utility's spokesmen had selected the one and only year among several recent years that might have been used to tell their story the way they wanted the commission to hear it.

When evaluating percent changes, it is always wise to ask yourself whether the source has an "axe to grind" and, if so, whether the base period used is unusual in some way that will ensure that the resulting percent change lends

---

[8] "Jones Urges McCracken to Correct Misleading Price Data," *Independent Petroleum Monthly*. XLI. No. 9 (January 1971), p. 5.

support to the source's biased position. If it appears that the base period has been handpicked with a view to supporting a bias, reject it. Easier said than done? Of course it is. But you might be surprised to see how skilled you become in spotting a phony percent change after just a little practice.

The Problem of Reducing Ratios

In Chapter 2, I commented on the advertising claim that nine out of ten doctors recommend the ingredients in a certain patent medicine. At that time our focus was on the use of recommendations for ingredients as a subtle proxy for recommendations for the product itself. Let us now worry about the first part of the claim, the part that says, "Nine out of ten doctors recommend...." What exactly does this tell us? The answer has to be nothing at all. To begin with, we are told nothing about the methods used to sample the doctors. Was the sample selected in a random manner? Were safeguards taken to minimize sampling error? Nonsampling errors? We simply don't know. Nor do we know the size of the sample—and here is where a common danger of ratios comes in. The sample could have consisted of only ten doctors. Or it could have contained ten thousand doctors or virtually any other number. If the sample contained only ten doctors and if nine indicated that they do recommend the ingredients in this product, then the ratio of affirmative answers to number of doctors responding would indeed be 9/10 (or nine-out-of-ten or 90 percent—they all say the same thing). If, on the other hand, ten thousand doctors were surveyed and nine thousand gave affirmative answers, then the ratio would be 9,000/10,000. Since grade school days, however, we have all followed the practice of reducing ratios as much as possible. In the present case, the ratio would become 9/10, the same as when the assumed sample size was only ten. The sample size might be very large or very small; the exact sample size has no bearing on the mathematics of reducing the ratio.

The sample size does, however, have a bearing upon how much trust we place in the 9/10 figure. From our point of view, the common practice of reducing such ratios destroys an important piece of information. Unless that piece of information is provided in some other way, one is well-advised to view the ratio with suspicion. A sample of only size ten would almost certainly be too small to represent adequately the views of the population of doctors with respect to the ingredients of the product under discussion (or just about any other subject). Since we are not assured that the sample consisted of substantially more than ten doctors, we cannot allow ourselves to place any confidence in the 9/10 figure. Even if we assume that this small sample was picked fairly, the 9/10 could still easily be a sampling fluke.

In the next chapter we consider what is, in my opinion, one of the most challenging families of common statistical fallacies to analyze—*Improper Comparisons.*

9

Improper Comparisons

Fate, Time, Occasion, Chance and Change?—To these all things are subject.

—Percy Bysshe Shelley

Making comparisons is an integral and important part of life. Some comparisons, such as whether Rembrandt's style of painting is more aesthetically pleasing than Van Gogh's or whether little Johnny requires more affection than his older brother, are best made on subjective grounds where statistics only get in the way. Other comparisons, however—such as whether a certain kind of foreign-built automobile really is more rugged than a similarly priced American make, or whether use of a new drug will help a patient recover faster than if nature were left to take its course, or whether the discount supermarket three miles from home really does charge lower prices than the one just down the street which gives trading stamps—comparisons such as these are aided by—indeed, dependent upon—statistics of some kind. But even in those situations where statistics can help, improper as well as proper comparisons are possible. In this chapter we examine several ways in which improper statistical comparisons are made.

The Dangling Comparison

We can dispose of one kind of *Improper Comparison* in a hurry—the comparison that isn't really a comparison at all.

An advertisement states that a certain brand of pipe will deliver smoke ten to twenty degrees cooler. Cooler than what? It never gets around to saying. The ad goes on to say that this pipe will produce up to 83 percent less tar and up to 71 percent less nicotine. Again, less than what? Another advertisement claims that a certain brand of tire will stop 25 percent quicker, but it never reveals what the tire will stop quicker than. Still another advertisement boasts that a particular brand of electric shaver has blades that are 78 percent sharper. Another says that a certain brand of denture paste will let you bite up to 35 percent harder without discomfort.

Examples could be listed *ad nauseum*, but little would be gained. Anyone who reflects even briefly on such claims realizes that even though precise-sounding statistics are bandied about, no real information is being conveyed.

Comparing Unlike Things

It isn't often that a statistical fallacy makes a little bit of legal history, but back in the early 1960's such was indeed the case. The advertising claims of National Bakers Services in behalf of Hollywood Bread were taken to task by the Federal Trade Commission on the grounds that they made a misleading statistical comparison. The claim made was that Hollywood Bread contains fewer calories per slice than other brands of standard commercial bread. However, the FTC maintained that Hollywood has as high a caloric content as any other bread—about 276 calories per 100 grams—and that the only reason a slice of Hollwood Bread contains fewer calories is that it is more thinly sliced; the comparison, therefore, was based on unequal units

WHO LIKES FAT BREAD?

of bread.[1] Or, expressing the same point in somewhat different words: like was not being compared with like.

Whenever two or more things are compared with respect to one characteristic, it is taken for granted that (1) the characteristic in question is the same for each, and (2) other relevant factors are close enough to the same that they won't invalidate the comparison. But comparisons are often invalid because one or the other part of this implicit assumption is not met. A recruiting sergeant, for example, might tell you that the death rate of American soldiers on a foreign battlefield is far less than the death rate in any big American city during the same period. What he will very likely fail to explain is that the city death rate includes the old, the handicapped, and infants, whereas the battlefield death rate is based on healthy young men between 18 and 35. Again, like is not being compared with like.

I recently heard a speaker state that the oral contraceptive is less dangerous than many leading drugs, including aspirin. "The number of aspirin poisonings make the pill look like a divine drug," he added. This speaker was obviously comparing the effects of the oral contraceptive taken in prescribed dosages with the effects of aspirin taken in excessive dosages. Any comparison of the harmful effects of two drugs should pit moderate dosage against moderate dosage or excessive dosage against excessive dosage; ie., like against like.

It has been pointed out that "Military prosecutors regularly run up an eye-catching 94% conviction rate (compared to 81% in the Federal courts for civilians)."[2] One should not hastily conclude, however, that military trials are little more than hanging courts. Many military trials are concerned with AWOL cases, whereas civilian courts do not often try people accused of not showing up for work.

In any given comparison, the assumption of sameness of the characteristic compared and of similarity of other relevant factors may or may not be valid. When it is not valid, the comparison is a lie. Sometimes much thought and maybe even research is required to determine whether the assumption is met. Such effort is essential if one aspires to avoid being taken in by arguments that fool a great many others—sometimes even the people presenting the arguments.

---

[1] Cited in Ray O. Werner, editor, "Legal Developments In Marketing," *Journal of Marketing*. XXVIII. No. 4 (October 1964), 78.

[2] *Newsweek*, August 31, 1970, p. 18.

Noncomparability of Definition
or Methods of Measurement

Did you know that at one time our southern states suffered a severe, prolonged malaria epidemic? That's right—in a manner of speaking. But the epidemic disappeared practically overnight. It seems that for many years prior to World War II hundreds of thousands of cases of malaria had been reported from below the Mason and Dixon Line, a situation which caused much concern when the building of southern training camps was begun in 1940. To protect its soldiers, the Army called in malaria experts from the Public Health Service who set out to locate proved cases and define the malarious areas. The upshot was that only a few dozen genuine cases were found. Most of the "malaria" wasn't malaria at all, not in the accepted medical sense. "Malaria" had been in the American South nothing more than a figure of speech used to denote cold or chill. When this regional peculiarity became recognized, the number of malaria cases plummeted from hundreds of thousands annually to only a dozen or so annually in the entire United States.[3] Any comparison made between number of malaria cases in the South before and after this discovery would, of course, be fallacious.

Comparisons are often made of unemployment rates for different countries. A difficulty with such comparisons is that the measures are based on different definitions of unemployment. As I pointed out in Chapter 2, the official definition of unemployment used in the United States is the most all-embracing of any country in the world, including as it does people on temporary layoff, people who are only available for part-time or temporary work, people with new jobs but who haven't yet reported for work, and so forth. Most foreign countries that gather unemployment statistics, on the other hand, count only those who register with employment exchanges.

Crime statistics showing percent changes over time in various crime categories have been widely criticized for reflecting changes in reporting methods along with actual changes in the incidence of specific crimes.

Mental and nervous disease appears more common in men than in women. Comparisons are tricky, however, because men are more likely to be detected and institutionalized, thereby getting their names on official records, since a higher proportion of them make their livings on jobs for which such disorders are likely to incapacitate them. An emotionally troubled woman

---

[3] Cited in Leonard Engel, "Danger: Medical Statistics At Work," *Harper's Magazine*, January 1953, pp. 79–80.

dependent on her husband's income would be less likely than her equally troubled husband to be included in a count of people with such illnesses.

Comparisons between two hospitals with respect to number of patients with specific severities of disease are often untrustworthy because of different standards of classification. If Hospital A assigns two-thirds of its patients to stages 1 and 2 and one-third to stages 3 and 4 while Hospital B assigns one-quarter of its patients to stages 1 and 2 and three-quarters to stages 3 and 4, do such figures indicate that patients actually present themselves to the hospitals in such widely differing proportions or are the differences due to different standards of classification?

Whether a comparison involves two or more readings on the same thing over time or readings on two or more things at the same point in time, definitions and methods of measurement must be identical or the comparison will be to some extent fallacious. What this implies is that a substantial amount of knowledge about what lies behind the figures may be necessary if one is to know whether a comparison is valid or contaminated by different definitions or different "mechanics" of data collection and organization. Again, some serious research may be necessary.

Comparing Percents When Actual Numerical Values Should be Used

In the preceding chapter I warned that percents can give quite a distorted impression if the base value used is extremely small. The presently relevant corollary of this point is: *When two or more things are to be compared, it is usually better to compare actual numerical values, if possible, then to compare percents if the base value of any of the percents is extremely small.*

The story is told about two men attempting to outboast each other over the educational attainments of their generations. "Forty percent of my high school graduating class went on to college," said one. The other countered with, "That's nothing; 50 percent of my high school graduating class went on to college. It would have been 100 percent except the other guy had to drop out of school for financial reasons." Clearly, with only two people in this particular graduating class, the statement including the 50-percent figure is much less instructive than a statement framed in terms of "one out of the two. . . ."

The Fallacy of the Sheep

An article in a New York newspaper some years ago stated that the chances of a married man becoming an alcoholic are double those of a bachelor, since figures show that 66 percent of "souses" are married men. The flaw in this argument is that roughly 75 percent of the men over 20 were married. So 75 percent of the men accounted for only 66 percent of the alcoholics—not such a bad record after all.

This example reminds me of the old story about the discovery being made that white sheep eat more than black sheep. Further investigation revealed why: There are more white sheep. Admittedly, the story isn't very funny, but it is uncommonly incisive. No small number of mistakes have been made over the centuries because people have overlooked the fact that there were originally more things in one category than in another.

The preceding point is probably too abstract. Let me attempt to clarify it using a hypothetical experience. Suppose that a certain Sunday morning was an especially grisly one for a particular city because six of its citizens were killed in separate traffic accidents. The breakdown according to destination was, let us say, as follows:

Going Fishing	Going to Church
4	2

Assume, for convenience, that fishing and church were the only two destinations for drivers that Sunday morning.

Accidents will happen (after all, two citizens got theirs while driving to church), but doesn't it seem a little odd that twice as many would-be fishermen were killed on this Sunday morning than would-be church attenders? Could it be that God really does mete out punishment to those who break the Sabbath? Before vowing to devote the rest of your life to church service, you would want to know the breakdown for the total population of drivers in that particular locale on that particular Sunday morning. Perhaps twice as many people on the road were going fishing than were going to church (a case, I suppose, of more black sheep than white sheep). In such a situation, it would not be too surprising to find that twice as many would-be fishermen were killed. If, on the other hand, only half as many were going fishing as were going to church, then going fishing on a Sunday might indeed appear inordinately risky. (Of course, some of the greater risk could be chalked up to the fact that the fishermen probably had greater distances to travel.) If, say, four times as many people on the road were going fishing as were going to church, then the risk associated with going fishing on a Sunday would appear slight relative to the risk of going to church. When trying to unravel this kind of comparison to see what, if anything, it suggests, we must know how many white sheep and how many black sheep there were to begin with.

An article about the frequency with which sharks attack people along American beaches asserted that statistics reveal that male victims in 1959 outnumbered females by 12 to one. The author asked rhetorically whether sharks, with their keen sense of smell, are attracted to something in the chemistry of men and repelled by women. Maybe. But before accepting this explanation, what other information would you want? You're on your own with this one.

While on unsavory subjects like accidents and death, I must mention a fallacy closely related to the Fallacy of the Sheep but which, in some respects, is more subtle. For want of a better label, I call it *the Risk-Your-Life-and-Live-Longer fallacy*, or simply RYLALL. RYLALL can be explained most conveniently through use of a familiar example.

The National Safety Council informs us that nearly half of the automobile fatalities occur at speeds of 40 miles per hour or less and that 65 percent of all accidents occur within 25 miles of home. Using these facts, much public service advertising points out that accidents are not solely the result of traveling at high speeds and are not limited to long trips. The conclusion drawn is that seat belts should be worn no matter how short the trip or slow the speed. So far, so good. I have no argument with this most sensible conclusion.

What such advertising doesn't tell us, however, is that most driving is done within 25 miles of home and at speeds of 40 miles per hour or less. No wonder there are more accidents under such conditions; there is more opportunity for accidents to occur. This omission could lead one to misconstrue such well-intentioned advertising and conclude that the wise thing to do is (1) always make certain you are more than 25 miles from home when driving, and (2) always drive at speeds in excess of 40 m.p.h. Granted, that would be

a pretty dumb conclusion to draw, but I am frequently surprised to find similar reasoning being applied to other situations. A report by the President's Commission on National Violence, for example, was widely interpreted in much the same bizarre manner:

Fact: Sixty-six percent of all rape and murder victims nationwide are friends or former friends or relatives of their assailants.

Conclusion reached in some news articles: You are safer in a public park among strangers at night than you are at home in your bed.

This conclusion is completely wrong despite the unimpeachable statistics. Everyone has friends or former friends or relatives and most people spend considerable amounts of time in close proximity with such acquaintances. Not many people, on the other hand, go for strolls in public parks during the dark hours in close proximity to no one but total strangers. It isn't surprising, therefore, that the majority of assailants are people with whom the victims are acquainted.

Several years ago, the claim was made in many newspapers that one is safer riding on the railroad than staying at home, thanks to the constant accident-prevention campaign carried on by the nation's transportation lines. This claim was usually accompanied by figures showing that fatalities in homes greatly outnumber railroad fatalities. However, such figures did not really substantiate the point being made. Since most people spend much more time at home than they do traveling on trains, this comparison was clearly improper.

Absence of a Comparison When One Is Needed

Sometimes a comparison is sorely needed to imbue statistical information with meaning. For example, an article in the *San Francisco Examiner* called "Church-goers Stay Wed" argued that church attendance is related to marital success or, more accurately, that nonattendance is related to marital failure.[4] The article contained the statement that a recent study had revealed that among 95 percent of the couples seeking divorce, either one or both partners do not attend church regularly. The statistic presented obviously does not support the conclusion implied by the title; the church-going or church-avoiding habits of divorcees-to-be tells nothing about such habits of those couples content to stay married. At the very least, we need assurance that some astonishingly high percentage of successfully married couples *do* attend church regularly. After all, maybe among 98 percent of the couples with demonstrably happy marriages one or both partners do not attend church regularly. I don't say that such is necessarily the case, but I do say that we have been provided too little information to be persuaded that there is any link between church attendance and marital success.

Another newspaper article stated: "Dr. _____ . . . says 60 to 75 percent of women who take birth control pills will suffer discomfort from contact lenses. . . . However, the problem will clear up as body chemistry adjusts to oral contraceptives, he told North Carolina optometrists."[5] The 60-to-70-percent figures would be more meaningful if we were told what percent of women who wear contact lenses but do not take the birth control pill suffer eye discomfort. If the figure were—let's make it really absurd—95 percent, then perhaps the pill should be prescribed for the relief of eye discomfort. Again, I don't say that anything like 95 percent is the true figure. In the absence of any facts or informed estimates, however, our imaginations are free to soar unrestrained.

In formal laboratory experiments the use of a standard for comparison in the form of a control group is considered essential. A treatment applied to one randomly assigned group—the so-called experimental group—cannot be adjudged successful unless the members of that group have changed significantly more than the members of the randomly assigned control group, the group not receiving the treatment. For example, suppose that biochemical researchers develop a new drug which they think might increase people's I.Q.'s. An experiment would probably be performed pretty much like the following: I.Q. tests will be given the group destined to receive the new drug for a trial period and also to the group destined to receive an ineffectual

---

[4] November 26, 1967.

[5] *The Denver Post*, December 3, 1969.

capsule, called a *placebo*, looking like the drug for the same trial period. No specific subject will know whether he is taking the real drug or the look-alike, and, perhaps, safeguards will even be taken to ensure that the researchers themselves don't know while the experiment is being conducted who is on the drug and who is on the placebo. At the end of the trial period, I.Q. tests will again be given all subjects involved in the experiment. If the average I.Q. of those receiving the real drug increases notably, a conclusion is still postponed until it is known what happened to the average I.Q. of the control group. If the control-group average is only about the same as before, the drug is assumed to have lived up to expectations. The control group is needed to ensure that the effects of other factors working on both groups are isolated form the true effects of the drug. What are some such other factors? Well, it is hard to say. Maybe I.Q. scores tend to improve with practice. Or perhaps the very act of participating in an experiment with such profound implications might induce subjects to try harder for high scores the second time the I.Q. tests are administered. There is even a remote possibility that some form of air pollution is insidiously raising I.Q.'s of the entire citizenry. Far-fetched though some of these possibilities may seem, the fact remains that many factors can affect test performance, and strenuous efforts must be made to ensure that the effects, if any, of the drug have been isolated from the effects of other factors.

Bradford Hill tells of an old but interesting case of a controlled experiment used to assess the value of a vaccination for the common cold. University students who believed themselves especially susceptible were invited to participate and were allocated at random to a vaccine group (the experimental group) and a water group (the control group). An extract from the results is shown in Table 6 below. Whatever the reason, the most impressive feature of

Table 6. Data for Cold-Vaccine Experiment

	Vaccine Group	*Control Group*
Number of persons vaccinated	272	276
Average number of colds per person		
Previous year	5.9	5.6
Current year	1.6	2.1
Percent reduction in current year	73	63

Source: Bradford Hill, *Principles of Medical Statistics* (New York: Oxford University Press, 1967), p. 276.

these figures is the eye-popping reduction in number of colds in both groups. We can readily imagine the kinds of faulty conclusions that might have been reached in the absence of a control group. What researchers could have helped but be impressed with a 73-percent reduction in colds? It is only alongside the reduction of 63-percent for the control group that the 73-percent falls into proper perspective and leads one away from concluding a cause-and-effect relationship between use of the vaccine and the reduction in incidence of colds. Dr. Hill summarizes the results of the experiment humorously in the following words:

> In fact, the vaccine, it appears, achieved nothing very remarkable for its recipients. Nevertheless, some of them were well satisfied, for the authors [who originally described this case in the *Journal of the American Medical Association* in 1938] report that from time to time physicians would write to them saying, "I have a patient who took your cold vaccine and got such splendid results that he wants to continue it. Will you be good enough to tell me what vaccine you are using? It must have been embarrassing to reply "water,". . . .[6]

Closely related to the topic of Improper Comparisons is that of *Jumping to Conclusions* from statistical evidence, the subject of the next chapter.

---

[6] *Principles of Medical Statistics* (New York: Lancet Ltd., 1967), pp. 276–77.

10

Jumping to Conclusions

As in your sort of mind, So in your sort of search; you'll find What you desire.

—ROBERT BROWNING

The *Newsweek* critic who reviewed a book, called *The Better Half*, about the early suffragettes ended his critique on a thought-provoking note. He wondered rhetorically what Susan B. Anthony and the other suffragettes would have said about the fact that almost 50 years after the enfranchisement of American women, a Columbia University sociologist found that only one wife in 22 said she casts a different vote from her husband.

A reader—a married man no doubt—wrote in saying:

> I feel that they would have been quite pleased. The feminist movement must have come a long way, if after fewer than 50 years since the enfranchisement of American women, only one husband out of 22 has the courage to vote against his wife.[1]

We find in this example seeming confirmation of an oft-harbored suspicion: Even when statistical data are above reproach, much leeway still exists for individuals to interpret them in a manner that best supports their prejudices. Or, as one cynic put it: "Statistics are like witnesses; you can get them to

[1] *Newsweek*, July 26, 1965, p. 2.

testify for either side." If you think that such dissimilar conclusions from the same data are rare occurrences, try following current economic statistics for a time. A report will be released to the effect that the Consumer Price Index increased at an annual rate, of, say, four percent during the preceding month. What happens? Spokesmen for the Administration herald it as proof that the President's anti-inflation policies are working according to plan. Political adversaries use the news as proof that the President's policies are accomplishing nothing. It happens all the time.

In this chapter I illustrate some ways in which a leap is made from a statistical fact to a convenient conclusion without the conclusion jumper being handicapped by any consideration of alternative conclusions or even by any rules of simple logic.

Hasty Conclusions

The story is told about physicist George Gamow who once worked in a seven-story building and often had to go from his office on the second floor to an office on the sixth floor. Curiously, whenever he wanted to take the elevator up, the first car to arrive was always going down. Putting his keen analytical mind to work on the matter, Gamow decided, "They must be making new elevator cars on the roof and sending them down to be stored in the basement." This explanation had its limitations, however, as Gamow soon realized, because whenever he wanted an elevator down from the sixth floor, the first one that stopped was almost always on the way up. So the physicist formulated a totally new theory: "Elevator cars are being built in the basement, then sent to the roof to be carried away by helicopters."[2]

Needless to say, there is an alternative explanation that is at least as plausible as Gamow's: If you are waiting on one of the building's lower floors, most of the elevator cars are likely to be somewhere above you. It is probable, therefore, that the first one to arrive will be headed down. Conversely, if you are on the floor near the top, most elevator cars are likely to be somewhere below you. Hence, the first to stop is probably headed up.

Gamow's explanations were, of course, offered with tongue in cheek, but it is only the intentional facetiousness that distinguishes this example from a great many others that could be cited where conulsions are arrived at in haste without regard to other equally plausible, or even more plausible, explanations. Before accepting any conclusions allegedly following from statistical evidence, a good practice is to ask yourself whether other, equally plausible, conclusions can be reached from the same evidence. What follows is a series of brief exercises of this kind. First the facts are presented. Then the conclusion drawn by the writer or speaker or whomever the conclusion jumper happens to be. Then an equally plausible alternative conclusion is offered by me or left for you to supply. Remember, the explanation offered

[2] Martin Gardner, "It's More Probable Than You Think," *Reader's Digest*, November 1967, pp. 107–8.

by me or by yourself might not be any closer to the truth than the one offered by the original source. They might both be wrong. Or they might both possess some truth. The point of this exercise is: *When equally plausible alternative conclusions can be reached from exactly the same statistical evidence, the logical link between evidence and conclusion offered is probably rather weak.* Someone just might be trying to soften you up with impressive-looking statistics before attempting to get you to believe his favorite preconceived notion. Or his motive might be perfectly innocent. He might not be aware himself that his conclusion is not the only one that can be reached from the data cited.

Fact: In 1952 just 44% of German males weighed 165 pounds or more. By 1963, 58% had reached or exceeded this weight."

Conclusion offered: Germany is becoming a land of fat people.[3]

Equally plausible alternative conclusion: Germany is becoming a land of people who are better nourished than was the case back in 1952.

Fact: "Since the end of 1958 the price of finished steel has increased 4% while the overall cost of living has risen 16%."

Conclusion offered: Steel price increases do not cause inflation.[4]

Equally plausible alternative conclusion: Steel price increases do not cause inflation single-handedly. Inflation is the result of many factors, steel price increases being only one contributory cause.

Fact: "Fifteen individual nations have improved their infant mortality rates more than the United States since 1950."

Conclusion offered: The state of health care in the United States is inadequate.[5]

Equally plausible alternative conclusion: The state of health care in several other countries—probably primarily underdeveloped countries—has improved markedly since 1950, at least with respect to infant mortality.

Some examples for you to ponder are presented below. Remember, your conclusion doesn't have to be any better than the one offered. It only has to be just as good and equally well (or, perhaps I should say, equally poorly) supported by the facts.

Fact: "In fiscal 1962, the Federal Trade Commission opened 1,795 formal investigations of suspected business abuses. Last year (1968) it opened only 611 . . . The FTC staff increased from 1,126 to 1,230 between 1962 and 1968."

---

[3] *The National Observer*, November 20, 1967.

[4] *Business Week*, December 9, 1967, p. 47.

[5] "Crisis in Health Care in the United States," *The International Teamster*, September 1971, p. 18.

Conclusion offered: The FTC is doing less work with more people. The overall conclusion of the article is that the FTC is an ineffectual commission.[6]

Fact: "The poor have twice as much illness; four times as much chronic illness; three times as much heart disease; seven times as many eye defects; five times as much mental retardation and nervous disorders."
Conclusion offered: Being poor is the cause of these various maladies because the poor can't afford the high cost of adequate health care.[7]

Fact: Only slightly over 20 percent of the population receive dental care each year.
Conclusion often offered: The dental profession is understaffed.

Drawing Conclusions About One Set of Items Based on Information About a Different Set

Sometimes the conclusion jumper will present valid factual information about one set of items and then proceed to draw a conclusion about quite a different set. The logical flaw is sometimes subtle. Without some serious thought you might feel convinced that the writer or speaker has proceeded quite logically from fact to conclusion. A good example of this kind of fallacy is found in this rather common argument regarding sex crimes:

Fact: Persons involved in illicit sexual activities each performance of which is punishable as a crime under the law constitute roughly 95 percent of the male population.
Conclusion sometimes reached: Only a relatively small proportion of males who are sent to penal institutions for sex offenses have been involved in behavior materially different from the behavior of most of the male population.
We have in the preceding a prime example of a statistical non sequitur, a Latin term meaning "It doesn't follow." Notice how in this example a fact is presented about the very large set of males who commit sex "crimes." (This set is indeed large because in most states many infrequently enforced laws do exist on the books which make a variety of common sexual practices— even between consenting adults and, in some instances, between married couples—strictly speaking, illegal.) But then a conclusion is drawn about the much smaller subset—those who are sent to prison for commiting illegal sexual acts. This group might consist primarily of rapists, child molesters,

[6] *Time*, September 26, 1969, p. 83.
[7] "Crisis in Health Care in the United States," *The International Teamster*, September 1971, p. 18.

exhibitionists, and what not, people whose sexual behavior does in fact differ considerably from the behavior of most males. The situation as it quite likely exists can be shown schematically as follows:

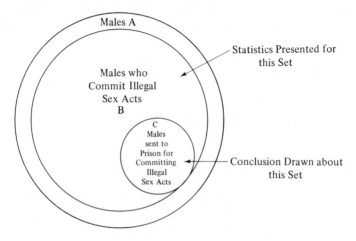

Figure 10–1

The members of the small set C may be quite a different breed from the majority of members in the larger set B.

In the preceding chapter I mentioned a newspaper article about the relationship between church attendance and divorce. The argument ran essentially as follows:

Fact: Among 95 percent of the couples seeking divorce, either one or both partners do not attend church regularly.

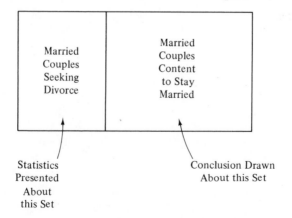

Figure 10–2

Conclusion offered: Church goers stay wed.

Again, the figure cited pertains to one set and the conclusion reached to a different set entirely. If such a thing is possible, the logic in this case is even more feeble than it was in the preceding example because the set about which the conclusion is drawn is not even a subset of the one about which the fact was given. (See Figure 10-2.)

Whenever facts are presented about one group of items, make certain that the conclusion reached pertains to the same group. When the groups are different, a logical error may not be absolutely inevitable, but the risk is great.

Now let us turn our attention to one of the most important foundation stones of formal statistical analysis—*probability*.

11

Faulty Thinking About Probability

> This branch of mathematics [probability] is the only one, I believe, in which good writers frequently get results entirely erroneous.
>
> —CHARLES SANDERS PIERCE

In Chapter 8 we dealt with percents in their role as descriptive measures. In this chapter we look at percents—or more accurately, ratios—in quite a different light, i.e., as probabilities. It is one thing to say that 75 percent of the people serving on a 12-person jury are male and quite a different thing to say: If one person on the jury were selected at random to serve as foreman, the probability is 0.75 that the person so selected would be male. Even though the figure 0.75 or 75 percent would be obtained by exactly the same mathematical process in the two instances—namely, by expressing the number of males as a ratio of the 12 people serving on the jury—there is an important difference in the way the 0.75 would be interpreted. When we report that 75 percent of those on the jury are male, we are merely describing the group in a particular way. Assuming our count is accurate, there is no uncertainty associated with the 75-percent figure. However, when we think in terms of selecting one person at random (that is, selecting him in such a way that chance alone determines who is chosen with personal judgment playing no role whatever), we are thinking about a situation where some uncertainty exists. We don't know prior to selection whether the person chosen will be male or female; we can only guess. Male would seemingly be a wise guess because the corresponding probability, 0.75, exceeds the 0.25 probability associated with selecting a female.

Probability is an important and exciting subject in its own right and as the foundation of inductive statistics, the branch of our art concerned with drawing conclusions about a total set of items from a subset or sample. Although I obviously can't turn you into an expert on probability theory and calculations, I do feel obligated to introduce you to some basic definitional and arithmetic concepts so that you can better understand the fallacies discussed.[1] Fallacious thinking about probability is exceedingly common and the errors made are often subtle and inseparable from the mathematical end of the subject.

The Statistical Fallacy
That Started It All

You might be interested to know that probability as a distinct branch of mathematics got its start from a statistical fallacy of sorts. A seventeenth century French nobleman and playboy, Antoine Gombauld, known as the Chevalier de Méré, raised certain questions about games of chance. To be sure, his interest in the subject was prompted more by practicality than by a thirst for knowledge for its own sake. You see, de Méré was a gambler who thought he had discovered a sure-fire winning strategy for betting on dice, and up to a point he was right.

De Méré had prospered for a time by betting that he could get at least one six in four rolls of a single die. He reasoned that this bet would win more often than it loses. He happened to be right but for the wrong reason. He figured that, since he had an even chance of rolling any one of the six numbers on the die on the first roll, the liklihood of a six is the same as for any of the other numbers, namely 1/6. For four rolls, he reasoned—and here is where the error creeps in—the chance would be four times as good. The probability of getting at least one six on the four rolls, therefore, should be 4/6 or 2/3. This reasoning led de Méré to conclude that he would, in the long run, win two wagers for every one he would lose. Actually the reasoning and the conclusion it leads to are wrong, but de Méré's fortunes were becoming progressively greater so his error went undetected. (The correct probability is about 0.52 in favor of his getting at least one six in four rolls. This figure is obtained by multiplying 5/6—the probability of getting something other than a six in a single roll—together four times to determine the probability

---

[1] You will find excellent discussions of probability theory in most orthodox statistics textbooks.

of his failing to get a six on all of the four rolls, and then subtracting the resulting product from 1.0.)

The wager that gave probability its start was a variation on the preceding one. For reasons unknown, de Méré switched to betting that within a sequence of 24 rolls of two dice, he could get at least one 12. Using the same line of reasoning as before, he decided that, because the probability of getting a 12 on one roll is 1/36 (there being 36 possible numbers that could be showing, only one of which is 12), in 24 rolls there must be a probability of 24/36 of getting at least one 12. In truth, however, this bet loses slightly more often than it wins, as de Méré's dwindling fortunes attested.

His perplexity led de Méré to write a letter to Blaise Pascal, mathematician and philosopher. In finding an answer to de Méré's puzzle, Pascal in conjunction with Pierre de Fermat, launched an inquiry into the theory of probability. From this unlikely start there gradually emerged an elegant and useful science. One can only conjecture about what might have happened (or, more accurately, failed to have happened) if de Méré's second wager had, been, by sheer dumb luck, a winner rather than a loser.

Ways of Looking at Probability

The word "probability" is rather tricky to define rigorously. Three somewhat different approaches to determining probabilities have been developed. These are summarized below.

A Priori Probability

Since probability theory had its origin in gambling games, it is not sur-prising that the method of measuring probabilities which was developed first was especially appropriate for gambling situations. The classical or so-called *a priori* concept defines the probability of an event as follows: If there are *a* possible outcomes favorable to the occurrence of event *A*, and *b* possible outcomes unfavorable to the occurrence of event *A*, and all the possible outcomes are equally likely and mutually exclusive, then the probability that *A* will occur is

$$\frac{a}{a+b} = \frac{\text{number of outcomes favorable to occurrence of event } A}{\text{total number of possible outcomes}}$$

The preceding probably sounds like jabberwocky if you haven't had any prior exposure to probability. But the concept is really quite simple and loaded with intuitive appeal. What all this gets around to saying is that if, for example, you have a coin believed to give heads and tails an equal chance of coming up on a fair flip, the probability of getting heads is obtained by first counting the sides of the coin. There is a head side and a tail side (the edge being ignored because of the extreme unliklihood that a coin would land on its edge and remain balanced there), and since that is all the sides available, we simply put down the figure 2 as the denominator of a fraction we are developing. The numerator is determined by counting the number of ways a "successful" result could occur, that is, the number of ways that heads could be obtained. Heads could be obtained by having the head side of the coin showing on top after the flip. Heads wouldn't be obtained if tails landed up. Therefore, there is only one way of getting heads and our fraction is 1/2. The term "mutually exclusive" is simply a fancy way of saying that if a coin lands with the head side uppermost, it cannot, as a result of the same flip, land with the tail side uppermost.

Similarly, we can compute the probability of getting a six with one roll of a single balanced die by counting the number of sides with the figure six on them, namely one, and expressing this as a ratio of the number of sides altogether, or six. So, the probability of getting a six in a single roll is $1/(1 + 5)$ or $1/6$.

To determine the probabilities in the above examples, no coins had to be flipped nor dice rolled. The probability calculations were based upon what we call logical prior (hence, a priori) reasoning—reasoning done without resort to any actual experimentation.

When determining probabilities with this approach, the probability of 0.0 is assigned an outcome impossible to obtain and the the probability of

1.0 is assigned to an outcome that is a dead certainty. Also, if the probability of an outcome is some figure between 0.0 and 1.0, the probability of not achieving that outcome will be 1.0 minus that figure, since the probabilities associated with so-called collectively exhaustive events must add to 1.0.

Relative Frequency of Occurrence

The a priori concept of probability, while useful in many situations, runs into considerable difficulty when dealing with certain types of questions. If one wishes to know the probability that at least one defective widget will be found in the next batch coming off an assembly line, a priori probability isn't much help. To be sure, we can say that the batch either will or will not contain at least one defective item, but the "will" and "will not" are not equally likely. Presumably, there are more batches without defective units than there are batches with one or more defective units. (The importance of the assumption of equally likely outcomes used in determining a priori probabilities should not be underestimated. I recently had a stock broker try to convince me that the probability that any given stock will decrease in value between two points in time is only 1/3 because there are only three possibilities—up, down, and no change. This kind of lame logic crops up with disconcerting frequency.)

Let us say, to cite another example, that you are curious about your chances of rolling a strike your next time at the line in bowling. It is true that you will either roll a strike or you won't roll one, but this knowledge isn't very helpful since strikes might be a rare occurrence for you—or a common occurrence. Again, a priori probability isn't much help. Sometimes resort must be made to empirical data.

Assume that you are given a coin to flip which you are told is biased—that is, either heads or tails comes up more frequently than 1/2 the time. Unfortunately, you are not told whether the coin is biased in favor of heads or tails. That is what you are required to figure out for yourself by flipping the coin.

So let's say that you flip the coin 1,000 times and find that heads comes up 800 times, tails only 200 times. This result could be stated as follows: The relative frequency of heads is 800/1000 equals 0.80. It certainly seems reasonable to assign a probability of 0.80 to the appearance of a head with this particular coin. On the other hand, if you had flipped the coin only five times rather than the 1,000 we have assumed, the resulting ratio of, say, 4/5 or 0.80 could not be accepted with much conviction as the true probability of getting a head on that particular coin.

This illustration alludes to a number of characteristics of the relative frequency approach to defining probability. Consider an experiment in which there are independently repeated trials. The number of outcomes a of an event A in which we are interested is recorded in n trials of the experiment. Then the relative frequency of the occurrence of A is a/n or the number of times a occurred during the specified number of trials expressed as a ratio of the number of trials altogether. It goes without saying that the greater the number of trials, the more confidence we have in assuming that our relative frequency calculation approximates the true probability of event A. If you were to flip this biased coin a million times with the resulting relative frequency of 0.78 for heads, then, even though it might seem like hair-splitting, you would be well-advised to conclude that the true probability of a head in a single flip is 0.78 rather than 0.80.

This concept of probability, as with any alternative concept, is not completely free of philosophical difficulties. Frequently the definition is presented in terms like "the relative frequency in the long run under a constant cause system." We have already seen what is meant by relative frequency. "Long run" means that a large number of trials is assumed. "Constant cause system" means that the many variables playing upon the experiment do not change their natures or relative degrees of influence from trial to trial. How well this last assumption holds up in any specific series of trials must often be a matter of judgment. In our coin-flipping example the assumption of a constant cause system is probably fairly safe assuming that precautions are taken to see that the coin gets a fair flip each time. On the other hand, determining the probability of your rolling a strike the next time at the line is not so simple. Not that the calculations are difficult; all you would have to do is examine your past record of rolls and express the number of strikes as a ratio to the number of frames you have bowled. But is the cause system constant? Probably not. Individual skills will improve, or in rare cases deteriorate, with repeated trials.

Subjective Probability

The subjective or personalistic concept of probability is of fairly recent origin. The probability of an event is the degree of belief or confidence placed in the occurrence of the event by a specific individual based on the evidence available to him. This evidence may consist of quantitative information or it may be totally qualitative. If the individual believes it is unlikely that an event will occur, he assigns to it a probability close to zero. If he believes the event is highly likely, he assigns a probability close to one. Subjective probability is a highly useful concept for certain kinds of problems, particularly those concerned with statistical decision making.

The Fallacy of
the Maturity of Chances

Many people seem to believe that there is something akin to the law of atonement in probability. If a coin thought to have no built-in bias produces nine heads in a row, they reason, one would be wise to bet on tails the next flip. After all, if the ratio of tails to total flips is 1/2 in the long run, then, in the aforementioned long run, there should be as many tails as heads and that means tails have some catching up to do. The fallacy in this line of reasoning lies in the fact that the coin has no memory nor conscience. The coin simply isn't aware that tails are lagging behind heads. Therefore, the probability of getting a tail on the next flip is still 1/2 just as it was prior to each preceding flip regardless of whether tails had been coming up about as often as heads or much less often.

The preceding is not to say that the long run ratio of 1/2 will not come to pass if the coin is flipped many, many times. What happens, however, is that the ratio of tails to total flips approaches 1/2 without necessitating a perfect balance between number of heads and number of tails. Let's consider a hypothetical case in order to show how this seeming anomaly works out. Suppose you flip a coin 100 times and get 60 heads and 40 tails. The discrepancy of 20, amounting to 1/5 of the number of flips, seems quite large. But if you flip the coin a thousand times with a discrepacy of 20 in favor of heads, this is now only 1/50 of the number of flips. With one million flips and a discrepancy of 20 in favor of heads, the ratio becomes an inconsequential 1/50,000. One could readily think up an example where the numerical difference between number of heads and number of tails actually increased as the number of flips increased without negating the fact that the ratio of tails to total flips tends toward 1/2. Don't give a coin, a pair of dice, a deck of cards, or any other nonhuman gambling device credit for having any independent horsesense. As sure as you do, your billfold will soon get that undernourished look.

This fallacy is heard frequently in quite different contexts as well. The contest entrant laments that she has never won anything before, so it is about her turn. Salesmen are urged not to get disheartened when a series of calls on potential customers fails to result in sales. The more losing calls one makes, so the reasoning goes, the closer he is to making a winning call.

Then, of course, there is the story about a surgeon who was about to perform a serious operation on a patient. Before the operation the patient came to him and asked to be told frankly what his chances were. "Why," replied the doctor, "your chances are perfect. Statistics show that 99 people out of 100 who have this operation die, but, lucky for you, my last 99 patients with this exact same malady have all died during the operation. Since you're the hundredth one, you haven't a thing to worry about."

A Fabulous Winning Strategy

Speaking of flipping coins, as we were just a few paragraphs back, this might be as good a place as any to call attention to a little different kind of gambling fallacy.

Books on betting sometimes describe, in all seriousness, this strategy for winning on a coin toss: Let the other person make the call because, if you're the one who makes the call, the chances are three to two against you. The explanation usually runs something like this: People call heads seven out of ten times, but heads will turn up only five times out of ten. Consequently, if you let your opponent do the calling, you have much better chance of winning.

This bit of nonsense sounds convincing until you think about if for a moment. Clearly, it doesn't make any difference who makes the call or what it is; the chances remain 50–50 every time.

Combining Probabilities

Many statistical fallacies occur when people try to take account of joint events. In this section I will show some right ways of handling joint events and then in the following section, some typical wrong ways.

Independent Joint Events

While we're still on this coin-flipping kick, let's consider the case of a flip of two coins. When we flip the two coins together, what happens to one coin has no effect on the other coin. Hence, we say that the events or outcomes are *independent*. When we try to compute the probability of getting, say, a head on each coin, we deal with so-called joint events, because we are considering the combined behavior of the two coins.

Okay. So now we know the terminology. How might we go about determining the probability of getting a head on each coin? Well, we know there are three possible end results

Head on both coins
One head and one tail
Tail on both coins

At first glance, it might seem as if these three possible outcomes are equally likely and that, as a result, the probability of both coins coming up heads is 1/3. But such is not the case.

In order to show most clearly the fallacy of such reasoning, I shall give the two coins individual identities by assuming that one is a dime and the other is a nickel. If we assume that the dime lands head-up, there are still two possibilities for the nickel, namely head-up or tail-up. Alternatively, if we assume that the dime lands tail-up, the same two possibilities hold for the nickel. Hence, we see that a complete list of possible outcomes would be

Dime	Nickel
H	H
H	T
T	H
T	T

Are these four possible outcomes equally likely? The answer has to be yes provided that the individual coins are not biased and our flip is fair. We see that only one of the joint events has both coins head-up, whereas two such joint events have head-up on one coin and tail-up on the other. By a priori reasoning, we see that the probability of getting two heads on the two coins is 1/4, or the ratio of number of favorable joint events to the total number of possible joint events. The probability of getting a head on the dime and a tail on the nickel is also 1/4. Note, however, that the probability of getting one head and one tail, no order specified, is 1/2 because there are two ways of being successful out of four possible outcomes.

What is the probability of having *at least* one head on the two coins? The denominator is still four. But the numerator would consist of the count of ways in which either one or two heads could be on top, namely three.

These probabilities could be computed more readily without the necessity of enumerating all the possible outcomes by employing the multiplication rule of probability which, for independent events, states: *If two events A and B are independent, the probability of getting A and B together or in succession is equal to the product of the separate probabilities.*[2] Using this rule, we can determine the probability of getting head on both the dime and the nickel simply by multiplying 1/2 times 1/2, which, of course, gives 1/4.

Determining the probability of one head and one tail on the two coins without regard to order is slightly more complicated. We know that the probability of getting head on the dime and tail on the nickel is 1/4. But the same probability would have to be attached to tail on the dime and head on the nickel. Therefore, the two 1/4's must be summed to get the correct answer, namely 1/2.

We can use similar reasoning when three coins—say, a dime, a nickel, and a penny—are flipped. The possible ways the coins can come up are

Dime	Nickel	Penny
H	H	H
H	H	T
H	T	H
T	H	H
T	T	H
T	H	T
H	T	T
T	T	T

The probability of getting head on all three coins is one in eight, or 1/8. Or, using the multiplication rule, we simply multiply 1/2 times 1/2 times 1/2 and arrive at the same answer. The probability of getting two heads and a tail on the three coins without any sequence being specified would be 3/8. The probability of getting two heads and one tail in any specified order would be 1/8, but there are three separate possible orders since the tail can appear on either the dime, the nickel, or the penny.

[2] In symbols:

$$P(A \cap B) = P(A) \cdot P(B),$$

where $P(A \cap B)$ is read "the probability of getting both A and B" and where $P(A)$ and $P(B)$ are the probabilities associated with events A and B, respectively.

This rule can be generalized to read

$$P(A \cap B \cap C \cap \cdots \cap N) = P(A) \cdot P(B) \cdot P(C) \cdot \cdots \cdot P(N)$$

Dependent Joint Events

Suppose we have a bag containing three red marbles and two blue marbles. If we should draw one marble from the bag and then replace it and draw again, we would have a case of independent joint events, the case we have just been considering. The probability of drawing two blue marbles is 2/5 times 2/5 equals 4/25. But if we do not replace the first marble drawn and proceed to draw again, the probability of drawing a blue marble is not 2/5 anymore. Assuming, as is necessary, that the first marble drawn is blue, the probability of getting the remaining blue marble on the next draw is only 1/4. (One blue marble remains among four marbles altogether.) The probability of getting two blue marbles, therefore, is conspicuously reduced, i.e., 2/5 times 1/4 equals 1/10, which, of course, is smaller than 4/25. Thus, when events are dependent, each fraction must be adjusted for the effect of previous events. For example, if we wished to know the probability of getting all three red balls in three random drawings without replacement, we would calculate

$$(3/5)(2/4)(1/3) = 1/10.$$

Of course, there are times when several such products would have to be summed. Suppose, for example, that we are curious to know the probability of getting one red marble and one blue marble in two drawings without replacement. The probability that the first marble drawn will be red and the second blue is $(3/5)(2/4)$ equals 3/10. But the requirement of the problem could be met by drawing a blue marble first and then a red, the probability of which would be $(2/5)(3/4)$, or 3/10. Note that both of these probabilities assume specified orders. Since either order, red first or blue first, will meet the requirements of the problem, the two probabilities must be added and the answer is 3/10 plus 3/10 equals 6/10, or 3/5.

Problems involving dependent events can be quite complex and I prefer to avoid any further encounter with the fundamentals of probability calculations. You should keep in mind, however, that the probability obtained when events are independent is different from that obtained when some kind of dependency exists. In short order I will present a true story of how two people were sentenced to lengthy prison terms because the prosecuting attorney in the case wasn't clear on the difference between independent and dependent events—nor was anyone else involved with the case.

Common Errors Made When Dealing
with Joint Events

As mentioned earlier, much fallacious thinking is done in connection with joint probabilities. I have found it convenient to classify such fallacies into three categories: (1) Carelessness, (2) Computing joint probabilities after some outcomes are known, and (3) Assuming events are independent when such might not be the case.

Carelessness

In an article in the *New York Herald Tribune* of May, 1954, it was stated that if the probability of knocking down an attacking airplane were 0.15 at each of five defense barriers (radar plus fighter planes), and if an attacking plane had to pass all five barriers to get to the target, then the probability of knocking down the plane before it passed all five barriers would be 0.75.

This is the same mistake de Méré made, i.e., determining the probability associated with one trial and then multiplying by the number of trials. The correct way of solving this problem would be to determine the probability that the plane will be shot down at the first barrier (0.15); then find the probability that it will get safely past the first barrier but be shot down at the second (0.85 × 0.15 = 0.1275); then the probability that it will get safely by the first two barriers but be shot down at the third (0.85 × 0.85 × 0.15 = 0.108375); and so forth until all five possibilities in which the plane could be downed have been considered; then sum the resulting products because any of the five possibilities means a downed plane. The probability resulting from this procedure is a nice round 0.5562946875. A much easier way of arriving at the same figure, however, is to find the probability that a plane will pass safely by all five barriers (0.85 × 0.85 × 0.85 × 0.85 × 0.85 = 0.4437053125) and subtract that product from 1.0.

In the book *The Bridge Over the River Kwai* there is an interesting but incorrectly executed exercise in probability. Three commandos are to be parachuted into the Siamese jungle for sabotage against Japanese forces. An RAF officer has been asked if, in default of the regular course, for which there is not enough time, he can give them some quick training in jumping under more adverse conditions. He attempts to discourage them saying:

> ... if they do only one jump, you know, there's a fifty per cent chance of an injury. Two jumps, it's eighty per cent. The third time, it's dead certain they won't get off scot free. You see? It's not a question of training, but the law of averages.[3]

There is apparently an error in these calculations. (I say "apparently" because, although it isn't mentioned, there is an outside chance that the probabilities given were obtained from empirical data which would have to be respected even if they fail to square exactly with the theoretical a priori calculations.) If the chance of coming safely to earth is 1/2 for one jump, it must be 1/2 times 1/2 for two, 1/2 times 1/2 times 1/2 for three. When we

---

[3] Reprinted by permission of the publisher, The Vanguard Press, from *The Bridge Over the River Kwai* by Pierre Boulle. Copyright, 1954, by Pierre Boulle.

subtract these products from 1.0, we find the probabilities of an injury from two jumps to be 0.75 and from three jumps, 0.875. By the way, in the story they all jump and none is hurt.

Computing Joint Probabilities After Some Outcomes Are Known

The story is told about a man who thought he could protect himself on plane trips by taking a harmless bomb along in his luggage. He reasoned that the odds against one person taking a bomb aboard were high, but the odds against two people doing it were surely astronomical.

A similar story concerns a sailor who put his head through a hole made in the side of a ship by an enemy ball with the intention of keeping it there for the duration of the battle. He explained that it was extremely unlikely that another ball would come in that exact same hole.

These stories are, of course, fictional and pretty far-fetched at that. Still, we hear the same kind of reasoning—reasoning akin to that used when committing the fallacy of maturity of chances—with depressing frequency. Some years ago in an article about a well-known golfer's sinking two holes-in-one back-to-back, the author said that the odds against making the first hole-in-one were huge, but the odds against sinking the second hole-in-one after just making one were astronomical. And, if you think the story of the cautious sailor is hard to believe, you might be surprised to learn that during World

War I, soldiers were encouraged to get into fresh shell holes because it was highly unlikely that two shells would hit the same spot during the same day.

The fallacy in all of these examples can most easily be explained through use of the multiplication rule and an example from a game of chance: What is the probability of getting the ace of spades twice in two successive drawings from a well-shuffled deck assuming that the first selection is placed back into the deck before the second drawing? Worded this way, the problem is pretty standard. We simply multiply the probability of getting ace of spades on the first drawing, 1/52 assuming jokers have been removed, by the probability of getting ace of spades on the second drawing—still 1/52. In other words, 1/52 times 1/52 equals 1/2704. But now suppose we have already drawn the ace of spades and have replaced it in the deck and are now contemplating the outcome of the second drawing. Since we know that we got the ace of spades the first time, the probability of that is now 1.0, indicating absolute certainty. The multiplication rule applied under these circumstances gives 1.0 × 1/52— exactly the same probability as that of getting an ace of spades in a single drawing.

The same reasoning applies to the other examples. The man who takes a bomb onto a plane, knows with certainty that he has it, hence he has not reduced the probability of there being another, more lethal, bomb aboard. The sailor who stuck his head through the hole in the side of the ship and the soldiers who got into fresh shell holes failed to realize that, even though it is improbable for a missile to hit on any specific spot, once it has hit, the probability associated with the known hit is 1.0, and, assuming independence of events, there is clearly no reduction in the probability that one will hit in the same spot again. Finally, although two holes-in-one back-to-back would be extremely unlikely when viewed before hand, once a hole-in-one has been made, that is now certain. Assuming independence of events, the next hole-in-one has the same probability associated with it as was the case before the other hole-in-one was sunk.

Assuming Events Are Independent When Such
Might Not Be the Case

In *Facts From Figures*, M. J. Moroney explains why every woman is one in a billion:

> Consider the case of a man who demands the simultaneous occurrence of many virtues of an unrelated nature in his young lady. Let us suppose that he insists on a Grecian nose, platinum-blond hair, eyes of odd colours, one blue and one brown, and, finally, a first class knowledge of statistics. What is the probability that the first lady he meets on the street will put ideas of marriage into his head? To answer the question we must know the probabilities for the several different demands. We shall suppose them to be as follows:
>
> Probability of lady with Grecian nose: 0.01
> Probability of lady with platinum-blond hair; 0.01
> Probability of lady with odd eyes: 0.001
> Probability of lady with first-class knowledge of statistics: 0.00001[4]

Upon multiplying these probabilities together, Moroney concludes that the probability that the first lady the man meets on the street will have such a combination of qualities is one in an English billion, that is, 0.000000000001. Moroney's reasoning is above reproach if the qualities on the list are actually independent as he assumes. I can fantasize a situation, however, where the characteristics are, to a degree, dependent. Suppose the woman is embarrassed about her odd eyes and tries to compensate by dying her hair platinum blonde. Finding that this doesn't help her overcome her embarrassment she becomes progressively more uncomfortable around people. Such a woman might well pursue an introvert kind of activity like statistics and, in due course, become quite excellent at it. In such a case, three of her traits would not be independent, and, consequently, the probability of finding such a woman might be drastically reduced to one in only a few hundred million. I am pettiflogging, of course, but the point has some merit. One must be quite certain that separate qualities are independent if he is to base decisions on the results obtained from applying the multiplication rule. The following remarkable case, is one where, at the outset at least, independence was assumed incorrectly:

> After an elderly woman was mugged in an alley in San Pedro, Calif., a witness saw a blonde girl with a ponytail run from the alley and jump into

[4] (Baltimore, Maryland: Penguin Books, Ltd., 1951), pp. 8–9. Copyright (C) M. J. Moroney, 1951.

a yellow car driven by a bearded Negro. Eventually tried for the crime, Janet and Malcolm Collins were faced with the circumstantial evidence that she was white, blonde, and wore a ponytail while her Negro husband owned a yellow car and wore a beard. The prosecution, impressed by the unusual nature and number of matching details, sought to persuade the jury by invoking a law rarely used in a courtroom—the mathematical law of statistical probability.

The jury was indeed persuaded, and ultimately convicted the Collinses (*Time*, Jan. 8, 1965). Small wonder. With the help of an expert witness from the mathematics department of a nearby college, the prosecutor explained that the probability of a set of events actually occurring is determined by multiplying together the probabilities of each of the events. Using what he considered "conservative" estimates (for example, that the chances of a car's being yellow were 1 in 10, the chances of a couple in a car being interracial 1 in 1,000), the prosecutor multiplied all the factors together and concluded that the odds were 1 in 12 million that any other couple shared the characteristics of the defendents.

Only one couple. The logic of it all seemed overwhelming, and few disciplines pay as much homage to logic as do the law and math. But neither works right with the wrong premises. Hearing an appeal of Malcolm Collins' conviction, the California Supreme Court recently turned up some serious defects, including the fact that not even the odds were all they seemed.

To begin with, the prosecution failed to supply evidence that "any of the individual probability factors listed were even roughly accurate." Moreover, the factors were not shown to be fully independent of one another as they must be to satisfy the mathematical law; the factor of a Negro with a beard, for instance, overlaps the possibility that the bearded Negro may be part of an interracial couple. The 12 million to 1 figure, therefore, was just "wild conjecture." In addition, there was not complete agreement among the witnesses about the characteristics in question. "No mathematical equation," added the court, "can prove beyond a reasonable doubt (1) that the guilty couple in fact possessed the characteristics described by the witnesses, or even (2) that only one couple possessing those distinctive characteristics could be found in the entire Los Angeles area."

Improbable Probability. To explain why, Judge Raymond Sullivan attached a four-page appendix to his opinion that carried the necessary math far beyond the relatively simple formula of probability. Judge Sullivan was willing to assume it was unlikely that such a couple as the one described existed. But since such a couple did exist—and the Collinses demonstrably did exist—there was a perfectly acceptable mathematical formula for determining the probability that another such couple existed. Using the formula and the prosecution's figure of 12 million, the judge demonstrated to his own satisfaction and that of five concurring justices that there was a 41% chance that at least one other couple in the area might satisfy the requirements.

"Undoubtedly," said Sullivan, "the jurors were unduly impressed by the mystique of the mathematical demonstration but were unable to assess its relevancy or value." Neither could the defense attorney have been expected to know of the sophisticated rebuttal available to them. Janet Collins is already out of jail, has broken parole and lit out for parts unknown. But Judge Sullivan concluded that Malcolm Collins who is still in prison at the California Conservation Center, had been subjected to "trial by mathematics" and was entitled to a reversal of his conviction. He could be tried again but the odds are against it.[5]

Some Unscrupulous Ways of Taking Advantage of the Fallacious Thinking of Others

Several years ago the following classified advertisement appeared in a magazine catering to believers in the supernatural:

Boy or girl?

What will baby be?

Famous psychic can tell from a snapshot or photo of full front view of mother-to-be. Must be after three months of pregnancy. Money back guarantee. Send $10.00 with photo to . . .

(Name and box number given)

Clearly, this famous psychic could guess right only half the time—no better than you or I could do—and still make $5.00 per guess, on the average. If he guesses right, he pockets the $10.00. If he guesses wrong he returns the $10.00, or so he promises. But it wasn't his money in the first place.

[5] *Time*, April 26, 1968, p. 41. Reprinted by permission from *Time*, The Weekly News Magazine; © Time Inc., 1968.

This is an example of one way that shysters can take advantage of others through probability. Another way is through sucker bets. As was demonstrated by the experience of Chevalier de Méré, described at the beginning of this chapter, the correct probability of an event or series of events is not always what common sense suggests. Shysters, as shysters will, have developed a number of wagers to exploit this particular failing of common sense. These wagers have a way of sounding as if the operator faces losing odds when in reality it is the pigeon who must lose in the long run. That is the reason why such wagers are labeled sucker bets. Because they illustrate ways that fallacious thinking is done about probability and because I would like to see you get reimbursed for the price of this book, I present a short sampling of such sucker bets in this section but request that you quit taking advantage of your fellow man just as soon as this book has been paid for.[6]

The Birthday Paradox

To employ this sucker bet, a gathering of thirty people or more (but preferably not too many more) is required. The operator says, I'll bet there are two people here with the same birthday." The pigeon, quickly calculating that the odds are about 11 to 1 against the occurrence of such an event, loses no time in taking the challenge. The pigeon has probably reasoned that there are 365 possible birthdays and only 30 people; hence, the probability is 30 in 365, or about one in 12. In truth, the odds are better than two-to-one that at least two of the thirty people will have the same birthday.

To see where common sense runs amuck, you might picture the 30 people lined up in a row. Number one states his birthday and the remaining 29 compare theirs with his. If there are no matches, number two announces his birthday and the remaining 28 compare their birthdays with that of number two. And so it goes. Each of the thirty persons has 29 separate chances of matching his birthday with another's. The typical pigeon thinks, "What are the odds that any one of the other 29 has the same birthday as mine?" He should be thinking, "What are the odds that any one of the 30 has the same birthday as any other one of the 30."

The Three Card Gambit

The operator shows the pigeon three cards. One is white on both sides; one is red on both sides; and one is white on one side and red on the other. He mixes the cards thoroughly in a hat and lets the pigeon select one at

---

[6] If you are a sucker-bet fan, you will find more examples in *Playboy's Party Games* (Chicago, Illinois: HMH Publishing Co., Inc., 1969), pp. 84–97. This little paperback also has an interesting section on "sure things."

random instructing him to place it flat on the table without looking at, or showing the operator, the under side.

If the upper side turns out to be red, for example, the operator will say something like, "It's obvious that this is not the white-white card. It must be one of the other two, so the reverse side can be either red or white. Even so, I'll give you a break. I'll bet you a dollar against your 75 cents that the other side is also red." It sounds like a fair bet. After all, doesn't the pigeon have a 50–50 chance of winning? Not on your life. The operator has a sure-fire money-making wager.

The hitch is that there are not two possible cases, as it appears, but three, and the three are equally probable. In one case the other side is white. In both the other cases, it is red, since the card selected from the hat can be either side of the red-red card. Therefore, the odds are two-to-one that the other side is red and the operator loses a dollar only half as often as he wins 75 cents.

The License Plate Ploy

In this one the operator offers to bet that one of the next ten cars that pass will have a double digit (11, 22, 33, etc.) as the last two numbers of its license plate. Although the bet sounds reasonable at even money, the chances of making good are about two-to-one in favor of the operator. One three-digit number in every ten has a double digit at the end and the operator is getting a full ten chances, not the five chances that would make it a fair bet.

The Eight Coin Con

The operator takes eight coins out of his pocket and asks the pigeon how many heads are likely to come up if he lets the pigeon flip all eight. The pigeon, aware that the odds of getting a head on any particular coin are 50–50, will undoubtedly say four. The operator then bets, giving two-to-one odds as a special enticement, that the pigeon can't flip four heads. The operator will make a sure profit in the long run. It is true, of course, that four heads is the average or expected value, but it will come up less often than the total of all other possible outcomes. The number of different outcomes is 256 of which only 70 are associated with exactly four heads.

Queens

In this one the operator takes two kings, two queens, and two jacks from a deck of cards, places them face down on a table, and shuffles them so that neither he nor the pigeon knows which is which. The operator then tells the pigeon that he, the pigeon, is so magnetic with the fair sex that if he picks two cards, at least one will be a queen. The odds are three-to-two that he has at least one queen. If the pigeon objects because there are six cards from which to choose, the operator offers him 10-to-one odds that he can't pick both queens. The operator still wins in the long run because the odds are 15-to-one against the pigeon's choosing both queens.

Sucker bets can be profitable just as long as you are the initiator. But when someone else takes the initiative it is well to heed the words of Sky Masterson, a character in Damon Runyon's "Guys and Dolls," who at one point in the musical relates the following tale:

> When I was a young man about to go out into the world, my father says to me a very valuable thing. He says to me like this: "one of these days in your travels, a guy is going to come up to you and show you a nice brand-new deck of cards on which the seal is not yet broken, and this guy is going to offer to bet you that he can make the jack of spades jump out of the deck and squirt cider in your ear. But, son, do not bet this man, for as sure as you stand there, you are going to wind up with an earful of cider."

Let us now turn our attention to *Faulty Induction*.

12

Faulty
Induction

A reasonable probability is the only certainty.

—E. W. Howe

Shortly before the presidential election of 1968 a widely read news magazine reported that, according to a poll conducted by a Wisconsin editor, George Wallace was favored over Richard Nixon by a ratio of five-to-three. But, the report continued, if Senator Eugene McCarthy had not withdrawn from the race, he would have been favored overwhelmingly. The magazine described these results as "puzzling."[1] And how, you ask, was the poll conducted? It seems the Wisconsin editor had questioned eight men in a bar! The only thing puzzling about the results is that they were considered worthy of space in a national magazine. Granted, this experience is not entirely unlike the sort of activity engaged in by some well-known and highly respected polling organizations. The Wisconsin editor questioned only part of the population of potential voters and then drew an inductive conclusion about the total population (or, at least, such was the insinuation of the magazine squib). It should be obvious, however, even to someone having had no prior experience with sampling, that a sample consisting of eight men in a particular bar is inadequate both in size and representativeness to reflect accurately the political views of members of the total population.

[1] *U.S. News and World Report*, October 21, 1968, p. 28.

The principal function of many formal statistical procedures is to help researchers make valid inferences about the aggregate of items of interest from a sample of such items. When certain rules are followed and precautions taken, such inferences can enjoy a high probability of being correct. Moreover, they can be most helpful in contributing to sound decisions with less cost, less loss of time, and maybe even greater accuracy than if a complete census were undertaken. But these rules and precautions must be observed religiously; otherwise, sampling only serves to delude the decision makers and others into thinking that a conclusion has been reached in a scientific manner. In this chapter we examine some of the main ingredients of valid statistical induction and several ways that statistical induction sometimes goes awry.

The Difference Between Induction and Deduction

The process by which one arrives at conclusions about an entire group of items from a study of particular cases is known as induction. Induction contrasts sharply with deduction which is not something you do at income tax time but rather is the process of drawing specialized conclusions from general propositions assumed to be true.

Most formal deductive reasoning depends upon so-called syllogisms wherein the reasoning runs from a major and a minor premise to a conclusion. For example,

Major Premise: All men like sports.
Minor Premise: Joe is a man.
Conclusion: Joe likes sports.

Induction, on the other hand, runs from the specific to the general:

Premise: Joe, Bill, Harry, Ike, Mac, et. al. are men.
Premise: Joe, Bill, Harry, Ike, Mac, et. al. like sports.
Conclusion: All men like sports.

The two modes of reasoning differ in a very fundamental way. In deduction, the conclusions are unquestionably valid if the major and minor premises are true. But, unfortunately, no new knowledge is acquired. The premises were already broader than the conclusion, a fact suggesting that one would have to possess a great amount of knowledge about the matter being considered even to set up true premises. (The major premise in my example, by the way, is certainly not true. Not all men like sports. Still, if you will play the game—that is, if you will humor me and grant that my premises *are* true—you are honor bound to accept my conclusion as well. Joe must like sports and that's all there is to it. Maybe it is this characteristic of deduction that prompted one cynic to define logic as "the art of going wrong with confidence and certainty.") Even though in deduction nothing new is revealed by the conclusion, the chain of reasoning can sometimes be quite helpful in bringing into view truths that are only implicit in the premises.

The conclusions in inductive arguments, on the other hand, are broader than the premises. If the conclusions are true, something has been added to the sum of human knowledge. But a price is paid for this possible extension of knowledge. Whereas a deductive conclusion must be true if the premises are true, an inductive conclusion might be quite untrue. There can be no absolute certainty about the conclusion even though one might be dead certain about the accuracy of the facts on which the chain of inductive reasoning is based. In short, invalid and downright misleading conclusions can be reached by the inductive process of reasoning.

Characteristics of the Inductive Process

Since our focus in this chapter will be on induction rather than deduction, certain characteristics of the inductive process should be elaborated upon. First, the conclusions from the inductive process hold only in terms of probabilities. Never can one be completely certain of his results. Induction by its very nature applies to cases not actually observed. When all items or instances are observed the procedure is not induction but enumeration.

Second, since a comprehensive account of all factors bearing upon an inductive argument is never possible, one who accepts the results of an inductive exercise (the results of a sampling survey, for example) places dependence on the knowledge and personal integrity of the ones presenting the information.

Third, to use induction we must assume that there exists some uniformity in the system of facts to which the premises and conclusion of inductive reasoning relate. This is the justification for the inductive leap from specific to general. We must be able to assume, for example, that a sample is a reasonably accurate reflection of the total population. (I plan to say more on this point shortly.)

Fourth, the verification of induction calls for reference to relevant facts in the real world rather than to the presence of internal consistency as is the case with deduction.

"Statistical inference" is a term used to denote the inductive process as it is used in connection with statistical data. When one gathers sample data and then attempts to make estimates of population characteristics or to test hypotheses about such characteristics, he moves inductively toward the drawing of a statistical inference.

How Valid Is the Inductive Leap?

Probably the most charitable way of viewing the example of the Wisconsin editor who interviewed eight men in a bar and then had his findings cited in a national magazine is to accept it as an indication of how far sampling has come in recent years in gaining public acceptance. Before the turn of the century most of the meager statistical data that were collected represented complete enumerations. Even in this century, formal sampling methods were used relatively little before World War II.

As recently as 1949 comedian Fred Allen made a disparaging remark about the Hooperatings (a radio listenership survey service) and, by implication, sampling in general. In so doing, he expressed the views of many who were still highly skeptical about sampling. Allen said: "The Hooperating is a so-called service that allegedly tells you approximately how many listeners the average radio program theoretically has. It's like taking a bite out of a roll and telling you how many poppy seeds there are in the entire country."[2] But there is one important flaw in the late comedian's analogy: To draw a

[2] Quoted in *Newsweek*, January 17, 1949, p. 48.

conclusion about how many poppy seeds there are in the country on the basis of one bite out of one roll would be to take a huge, extremely hazardous inductive leap from particular to general. To conclude that a certain radio show has approximately *X* number of viewers, on the other hand, would be much less hazardous assuming the sample size was reasonably large and the sample respondents were selected in such a way as to be representative of the population of radio listeners—which brings us to the subject of this section, namely the validity of the inductive leap.

So far, we have considered only the general structure of an inductive chain of reasoning. Whether the chain of reasoning is good or bad, right or wrong, all inductive arguments contain (1) a collection of facts about specific items or instances, and (2) a conclusion asserting that what is true of the few is also true of the many. Now we come to the practical question of how one tells whether an inductively drawn conclusion has a reasonable chance of being true or whether it is quite likely untrue. Let us consider four instances close to every-day experience:

Example 1: A doctor extracts a small amount of blood from you and concludes that your blood sugar is normal.

Example 2: A lady shopper buys a cup of strawberries on the basis of close inspection of eight individual strawberries in the cup.

Example 3: A quality-control engineer inspects 10 cleavis springs out of a batch of 25 and concludes that a sufficiently large proportion of all the springs in the batch are satisfactory.

Example 4: A young man has been jilted by three girl friends all under five feet tall and vows never to date short girls again.

Now pretend that you are asked to evaluate the quality of each of the above inductive conclusions. How do you proceed? You probably begin by recognizing that two steps in evaluating each such conclusion can be discerned: First, you must get the argument clearly in mind. More specifically, you must distinguish between the sample of items actually observed, the group generalized *from*, and the population of items, the group being generalized *to*. In Chapter 10, I warned you to be wary about those situations where information is presented about one set of items and a conclusion is drawn about a different set. But we do this very thing when we reason from sample to population, except that in this latter case, one set is a subset of the other. If the subset is representative of the larger set, then the inference stands a pretty good chance of being true. If it isn't representative, then the inference runs a high risk of being wrong—which brings us to the second step in evaluating an inductively drawn conclusion, namely making up your mind about it. This step involves the determination of whether the sample constitutes adequate evidence of the inference drawn. That is, you must ask yourself whether there is any reason to think that the sample might be biased or misleading. Let us keep these two steps in mind as we consider the four hypothetical cases listed above.

Example 1: Normal Blood Sugar Concluded
from a Small Sample of Blood

You begin your assessment of this conclusion by distinguishing clearly between sample and population. In this case the task is easy. The sample is the small amount of blood that the doctor extracted from you. The population is all the blood in your body at that particular point in time. So far, so good. Now you ask yourself whether the sample might be biased or misleading. You decide that you may safely assume, as does the doctor, that your blood is uniform throughout your body. Hence, anything that might be said about the small amount of blood can safely be assumed to hold for the larger amount.

Example 2: A Cup Of Strawberries Is Bought
on the Basis of a Close Inspection
of Eight Berries in the Cup

Distinguishing between sample and population is easy in this example too. The sample is the eight strawberries selected for close scrutiny and the population is all the strawberries in that specific cup. Now, you ask yourself, is there any reason to assume that the eight strawberries constitute a biased or misleading source of evidence about all the ones in the cup. The answer, you quickly realize, is maybe yes and maybe no, depending on how the eight strawberries were selected. If the sample were a so-called convenience sample consisting of eight strawberries found on the top of the cup, you would have good reason to doubt its representativeness. Storekeepers and berry packers are known to put the best berries on top. A careful shopper does not judge the whole cup by the berries on top; instead she pours a few out into her hand in order to see what the ones underneath look like. To summarize: You can't really judge the adequacy of the inference drawn in this case since you have no information which can help you evaluate the representativeness of the eight sample berries.

While on this subject, I must warn you about this overworked word "representativeness." It is a good, most informative word when used with judicious restraint. But it is also a word that is used more often in connection with real-world samples than the realities merit. Let us suppose that you watch our hypothetical shopper inspect eight strawberries on the top of the cup and conclude that all the strawberries in the cup are satisfactory. She then informs you that she gave her approval to that particular cup on the basis of a representative sample of berries from it. What choice would you have other than to conclude that she doesn't really know what a representative sample is? If she really knew what the term implies, she would hardly use it so loosely. Unfortunately, we have in this hypothetical situation something very close to the way things often work in real life. A sample is labeled "representative" whether or not there is any sound basis for believing it is. The story is told about a southerner who was thirty years old before he realized that "damn yankees" is two words. I have a sneaking suspicion that many people don't realize that "representative sample" is two words. The adjective "representative" gets tacked onto the noun "sample" with such carefree abandon that the term almost seems like a single word. Just remember it is very easy for someone to refer to a sample as representative whether or not it actually is, and writers of newspaper and magazine articles and research reports do just that with appalling frequency. But for researchers to obtain a truly representative sample is much more difficult. It is always a good idea to postpone acceptance of the claim that a specific sample is representative of the total population until you know more about how it was selected.

Example 3: A Batch of Cleavis Springs Is Accepted on the Basis of a Sample of Ten

We have in this example a situation much like the previous one. To begin with, distinguishing between sample and population is quite simple. The sample is the ten items actually inspected; the population is the batch of 25 such items. But are the ten items representative of the 25? As with the preceding example, you would want to know how the sample was selected. In this case, however, the issue of representativeness might be a good deal more clouded. You are not completely certain whether a convenience sample, consisting of, say, the ten cleavis springs within easiest reach would necessarily represent biased or misleading material. Much depends on a number of extremely subtle factors including the order in which the items are turned out by the production process and any unconscious biases that the inspector might harbor. If there are no such subtle biases present, a convenience sample might be just as representative of the whole batch as a sample selected in some other way. Just to be on the safe side, however, the inspector would be well-advised to take the precaution of selecting the ten sample items through a random procedure. Random sampling means that the sample items are selected without individual judgment playing a role. (I'll say more about this point in the next section.) Random selection can be facilitated through use of prepared tables of random digits available in many statistics textbooks. It would take the inspector only a small amount of extra time to select the sample items at random and the payoff in terms of objectivity might be well worth it.

Example 4: A Young Man Rejects All Girls
Under Five Feet Tall as Possible Dating Partners
Because Three Such Short Girls Have Jilted Him

In this case, although the sample is easy to identify, being made up of the three girls under five feet in height who have jilted this embittered young man, the population is much more vague than in the previous examples. Presumably, the population consists of all eligible girls under five feet tall whom this young man might conceivably date if he were aware of their existence and if external conditions permitted.

You would have good reason to doubt whether the three girls in the sample were representative of all the girls in this nebulous population. It goes without saying that people, including short girl friends, differ a great deal from one another. That is, they are highly variable with respect to many important characteristics. A sample of only three short girls would certainly be much too small to reflect accurately the substantial variation in personality traits in the population being judged.

Some Fundamentals of Sampling Theory and Practice

Throughout this book I have assumed that your goal is to learn to evaluate with some sophistication statistical evidence presented by others. I have not assumed that you will be doing any elaborate statistical analysis yourself (although you may, and I hope you do). But if you are to evaluate the worth of information obtained from samples and the inductive conclusions derived therefrom, you must be familiar with at least a handful of basic sampling concepts, terms, and procedures. In this section I shall touch briefly on the bare minimum collection of sampling topics with which you should be quite familiar.

An Important Basic Assumption

The single most important sampling concept is this one: If sample items are chosen at random from the total population, the sample will tend to have the same characteristics, in approximately the same proportion, as the entire population. Notice the emphasis on random selection. If the sample really is selected in a random manner, we can place great confidence in this basic assumption provided we show proper respect for the word "tend." Sample characteristics only *tend* toward or approximate the corresponding population characteristics.

Probability Samples, Judgment Samples,
and Convenience Samples

Three fundamentally different approaches to sampling can be distinguished. These are probability sampling, judgment sampling, and convenience sampling. Statisticians, generally speaking, have relatively little to say about the last two approaches save for warning of their inherent limitations. Probability sampling, on the other hand, is the cornerstone of much formal statistical analysis wherein information about a sample is utilized to make precise guesses about the corresponding population.

The unique characteristic of all probability sampling procedures is that the selection of items from the population for inclusion in the sample is made according to known probabilities. This characteristic of probability sampling implies three other features: (1) A specific statistical design is followed, (2) The selection of items from the population is determined solely according to known probabilities by means of a random mechanism, usually a table of random digits (to be discussed shortly), and (3) The sampling error—that is, the difference between the result obtained from a sample survey and that which would have been obtained from a census of the entire population conducted using the same procedures as in the sample survey—can be estimated and, as a result, the precision of the sample result can be evaluated. Notice particularly that with probability sampling, personal judgment about which population items should be included in the sample is ruled out. Moreover, once a sample item has been selected using the random mechanism, it must be included in the sample and not arbitrarily discarded.

So what are judgment and convenience samples? These terms must be touched on even though the main focus of the remainder of this section will be probability sampling. Sometimes a judgment or a convenience sample will be presented as if it were a probability sample. For that reason alone, the statistical critic should know the difference.

In a judgment sample, personal judgment plays the key role in determining which population items are selected for inclusion in the sample. The selection of "representative" sample items is a matter of personal conviction rather than the outcome of an impersonal random mechanism.

A convenience sample is merely a part of the population that happens to be conveniently at hand. A former marketing professor of mine, for example, one day brought some disposable paper ashtrays to class for the purpose of soliciting the opinions of the class members regarding the probable usefulness and popularity of this idea. He was obviously utilizing a convenience sample.

Judgment and convenience samples have legitimate places in research work. But they do lack certain advantages of probability sampling, namely lack of bias in selection and amenability to measurement of sampling error, and should not be passed off as true probability samples.

Before leaving this last point, I must emphasize that a convenience sample bearing only the most superficial possible resemblance to a probability sample is sometimes passed off as the latter. Reference to a fairly recent television commercial should help to clarify this point. In this commercial, a man runs out into the street and asks each of three people in cars—apparently waiting for the light to change—whether he or she uses a certain product. Two answer "yes," and one answers "no." The viewer, unless he is on his guard, not only tucks away the impression that two out of three people use this product but also that the sample is random. The interviewer picks out three cars quite arbitrarily, hence, the seeming similarity to random sampling. But this is still a convenience sample and not a random one. A truly random sample would give the stay-at-homes and people in other parts of the city a chance to be included as well as drivers in that particular neighborhood.

In an article called "What Makes A Perfect Husband?" author Sam Blum states:

> I interviewed, at random and with relish, secretaries, college girls, school-teachers, a young woman standing in line to see a movie, two dancers, a salesgirl, several women temporarily between marriages, the wives of friends —a whole array of assorted husband watchers.[3]

[3] "What Makes the Perfect Husband?" *McCall's*, August 1967, p. 61.

I don't doubt that author Blum did his interviewing with relish, but I can't help doubting whether he really did it at random. It sounds more as if he sampled the ladies haphazardly which, to repeat a point, is not the same as doing it randomly.

Simple Random Sampling

There are several different kinds of probability sampling techniques, the simplest to discuss being simple random sampling. The definition of a simple random sample is that it is a sampling procedure which ensures that each possible different sample of a given size contained within the population has an equal chance of being selected. And how does one go about seeing that this condition is met in practice? What follows is a brief summary of the physical procedures involved in selecting a simple random sample:

Let us refer once more to our quality-control engineer facing a batch of newly produced cleavis springs and wishing to assess their quality by examining only ten springs in the batch. To select a simple random sample of size ten, he would have to begin by giving each of the springs in the batch a separate identity. Perhaps he would simply line them up and number them 01, 02, 03, . . . , 25. Then he would have to decide how to go about selecting the ten needed sample items using a table of random digits. A table of random digits is exactly what it sounds like, i.e., a table of figures which were themselves generated by a random mechanism. Table 7 shows a hypothetical excerpt from such a table.

Table 7. Portion of a Table of Random Digits

13	41	34	<u>15</u>	20	24	03	65
<u>21</u>	40	70	42	38	94	88	28
99	<u>24</u>	87	08	49	66	98	32
00	95	93	60	84	87	09	10
60	80	83	73	78	<u>07</u>	66	31
28	01	74	19	27	67	66	00
22	03	30	00	83	64	94	41
<u>05</u>	78	81	56	33	27	<u>04</u>	65
19	09	98	33	52	89	87	69
57	64	28	33	67	93	92	98
36	24	72	78	39	58	20	90
70	99	46	69	97	44	72	36
99	63	89	47	14	63	18	67
72	05	51	57	38	57	99	43
24	03	86	55	57	88	55	<u>14</u>

Source: Adapted from an actual table of random digits.

Any way our inspector decides to use the table is legitimate provided he makes all decisions regarding its use in advance and then sticks religiously to his self-imposed rules. Let us say he somehow decides to start with the left-most pair of digits in the second line of the table and then selects every second figure as he proceeds down the column. Using this approach, he finds that the first number obtained is 21. Therefore, item 21 in the lineup will be included in the sample. The next figure, 00, cannot be used (the lowest possible usable figure being 01), so he ignores it. Next comes 28. This one is too large (there being only 25 items in the entire batch) so he ignores this one as well. The next one is 05. Hence, item number five will be included in the sample. He continues to use this procedure (ignoring all figures smaller than 01 and greater than 25 and any duplicates obtained) and finds that his sample is made up of cleavis spring numbers 21, 05, 24, 03, 09, 15, 08, 07, 04, and 14. This sample of ten items would meet the definitional requirement of a simple random sample not because of any characteristic of the items selected but by virtue of the manner in which the sample was chosen.

Other Kinds of Probability Samples

Although simple random sampling is conceptually the simplest form of probability sampling, it is not used a great deal in practice, especially where human populations are concerned, because of practical problems difficult to overcome. What if the relevant population consisted of all the people in the United States, as is often the case? The impossibility of getting an accurate, up-to-date list of all members of this huge population hardly requires elaboration. Even if one could obtain such a list, the mere act of numbering well over 200,000,000 names would be an exorbitantly expensive and time-consuming job. Moreover, attempting to make contact with the sample members selected could pose problems difficult, if not impossible, to overcome. One member of the sample might be a resident of San Francisco's China Town, another the owner of a large spread in Wyoming, another a fisherman in Maine, another a member of the jet set who spends most of his time in foreign countries, and still another a miner in the Appalachia region, and so on. Individual respondents chosen at random might turn out to be very nearly inaccessible themselves and be separated from one another by hundreds of miles. Another disadvantage of simple random sampling is more theoretical. Simple random sampling isn't as efficient in terms of minimizing the sampling error as are some alternative probability sampling designs. For these reasons, other kinds of probability sampling procedures have been developed. I shall touch on three of these briefly.

Stratified Random Sampling

In stratified random sampling, the population is classified into mutually exclusive subgroups or strata, and simple random samples are drawn independently from these subgroups. The primary purposes of stratification are to ensure representativeness and to effect a reduction in the sampling error by grouping together sample items which are more alike with respect to the characteristic under investigation than are the population items as a whole.

Cluster Sampling

Cluster sampling is a technique in which the population is subdivided into groups or "clusters" and then a probability sample of these clusters is drawn and studied. For example, if our population were all the households in the United States, we might draw a simple random sample of counties in the United States, and then draw a simple random sample of city blocks within the selected counties.

Systematic Sampling

In a systematic sample every *Kth* item is drawn from a population listed or arranged in some specific order, often alphabetically. The starting point is selected at random from the first *K* items. This is a convenient kind of probability sampling procedure when a list already exists and there is no reason to believe that a rhythmn of some kind is present in the list.

There are other kinds of probability sampling designs as well, but for the statistical critic, probably the main point of all this for you to keep in mind is that with any kind of probability sampling procedure a random mechanism, rather than personal judgment or convenience, is used to obtain the sample items. Don't let anyone try to pass of a judgment sample or a haphazardly drawn convenience sample as a true probability sample.

The Question of Sample Size

How large should the sample be? This is an important question, but, unfortunately, one quite impossible to answer in general terms. Sample size depends on such theoretical considerations as (1) amount of variation in the population (or the rarity of a specified event), (2) degree of precision required, and (3) the kind of sample design used as well as on such practical considerations as amount of manpower, money, and time available. A detailed discussion of how to determine an appropriate sample size is better left to specialized books on sampling. Two important facts for a statistical critic to note, however, when sizing up evidence obtained from a sample, are (1) the nature of the population with respect to variability (or rarity of a specified event), and (2) the sample size actually used. The more variable the population or the rarer a specified event, the larger the sample size should be, other things being equal. Since the chemical composition of a person's blood doesn't vary a great deal, a relatively small amount of blood serves as a perfectly good sample. Maybe eight randomly selected strawberries in a cup or ten randomly selected cleavis springs in a batch of 25 is satisfactory also. On the other hand, many kinds of human characteristics and attitudes vary a great deal—which is precisely why more than three girls under five feet tall would be required before passing judgment on the entire population and why more than eight men in a bar would be needed to determine the political leanings of an entire nation.

Examples of Dubious Sampling

Undoubtedly the most widely publicized sampling fiasco in the relatively short history of the art is the one concerning the *Literary Digest's* error in predicting the winner of the presidential election of 1936. The *Literary Digest*, which, incidentally, ceased publication shortly after its history making mistake, mailed out 10,000,000 ballots and had 2,300,000 returned. On the basis of this unusually large sample, it confidently predicted that Alfred M. Landon would win by a comfortable margin. As it turned out, however, Franklin D. Roosevelt received 60 percent of the votes cast, a proportion representing one of the largest majorities in American presidential history. A major difficulty with the sample was that the people to whom the magazine's ballots were sent were primarily upper-income types. This fact was the result of the magazine's having selected names from lists of its own subscribers, and telephone and automobile owners. These lists tended to be biased in favor of higher income groups, a most strategic kind of bias as it turned out. In the 1936 election there was a strong relationship between income and party preference. In the four previous elections, the *Literary Digest*, basing its predictions on exactly the same kind of samples, correctly predicted the winners. But in these elections there was much less of a relationship between income and party preference. In this election, the presence of bias rendered the sample, despite its enormous size, inadequate as a representative subset of the population.

A few years ago it was reported as a result of a sample study that one-sixth of the nation is ill-fed. This conclusion, based on research conducted by a medical doctor affiliated with the U.S. Public Health Service, is of questionable worth because the sample was not representative of the entire nation. The report stated that of 12,000 people examined, mostly in—and this is important—Texas, Lousiana, and Kentucky plus several hundred from upstate New York, 17 percent were undernourished enough to represent "real medical risks."[4] The conclusion drawn may or may not be close to the truth. The reliability of the sample as a representative subgroup of the population, however, is highly questionable (to state the case as politely as I know how). Clearly, the population actually sampled and the population for which the inference was drawn are two quite different entities.

In an article emphasizing the point that smokers breathing polluted air are much more susceptible to lung cancer than nonsmokers, the following statement was made:

> For example, a recent study of 370 asbestos workers showed not one case of lung cancer among the non-smokers, whereas lung cancer among the smokers was much more prevalent than would be expected in a normal male smoking population. The explanation given by scientists to account for this increased incidence is the presence of chemicals in the contaminated atmosphere, superimposed upon the damage done to the lungs by cigarette smoke.[5]

The point made in this article might be valid enough. However, I doubt whether the study described does much to support it. Three hundred and seventy people, divided into two groups, smokers and nonsmokers, seems like a pretty small sample in view of the relative rarity of lung cancer.

"A study comparing 34 drivers who had taken courses with 466 who had not, revealed that the driver-educated graduates actually had an accident rate 15% higher than their untrained counterparts."[6] Again, the sample sizes are probably too small, especially the one for drivers who had taken courses, to justify the inference drawn.

A few years ago an experiment was described in which two scientists put addict volunteers on large daily doses of methadone to block their craving for narcotics. The results were considered successful in that more than half of the patients under treatment for three to six months held jobs, and eight out of ten who had taken methadone for a year held jobs. According to the

---

[4] *Time*, January 31, 1969, p. 74.

[5] Gerhard Angermann, "More Bad News for Smokers," *Reader's Digest*, February 1969, p. 94. (Condensed from *Air Pollution Control Progress*.)

[6] *Time*, November 3, 1967, p. 49.

scientists conducting the study; "We can now define a treatment program that will work for a very large number of street addicts."[7]

I don't include this example in a section on dubious sampling with a view to suggesting that the experiment could have been conducted much differently. The scientists in this case had no choice but to work with addict volunteers. Still, one must keep in mind when interpreting the findings of a study in which volunteers where used that the sample might be quite biased if all addicts are to be construed as the relevant population. The type of person who volunteers for such a study and the type who refrains from volunteering might be very different. Before drawing inductive conclusions, one would presumably have to define the population in a rather restrictive and nebulous way. The population is certainly not all addicts but rather all addicts with a state of mind, or motivational system, or whatever you prefer to call it, similar to the addict volunteers in the study.

Overlooking Sampling Error

Earlier I stated that an inductive conclusion can legitimately be stated only in terms of probability, absolute certainty being impossible. In a situation where a sample characteristic has been measured and used to estimate the corresponding characteristic of the population, the results, ideally, should be stated with a so-called confidence interval and a probability statement indicating the degree of confidence one can have in accepting that interval as accurate. I am alluding to a technical area but would prefer not handling the point in a technical way. A more fruitful approach might be to present an example and examine some of its failings.

The following summary of a Gallup poll appeared in many newspapers across the country:

> A majority of American adults (6 in 10) would advise young people today to take up teaching as a lifetime career, with the underlying belief that today as never before society needs dedicated people to deal with the nation's youth.
>
> This question was asked: "If you were advising a young person on choosing a career, would you advise this person to become a teacher or not?"
> The results:
>
> Yes . . . . .62 per cent
> No . . . . .24 per cent
> No opinion. . . . . .14 per cent[8]

This article made no mention of the important fact that the percentages represented the responses of only several hundred adult Americans—not *all*

[7] *Newsweek*, October 10, 1966, p. 77.

[8] This specific version of the report came from *The Denver Post*, November 20, 1968.

adult Americans. True, the known sample percentage and the unknown population percentage should be rather close if the sample was selected in an unbiased manner—and I feel quite confident that the Gallup organization would have seen to that. Nevertheless, there will be discrepancies between sample and population percents because the sample contains only some, not all, population items and because we cannot safely assume that this specific sample (or any other sample that might have been selected) is an exact minia-ture replica of the parent population. The 62 percent answering "yes," for example, might have been, say, only 58 percent had a complete census rather than a sample been used. Or it might have been 65 percent or some other figure somewhere in the general vicinity of 62. Although because of the lack of statistical sophistication of a large proportion of the general public it is not likely to happen soon, the way such information should be presented would be with a confidence interval—instead of 62, for example, a range of figures like 58 to 66, or whatever, would be shown—and a probability figure indicating that we can be 90 percent confident, 95 percent confident, 99 percent confident, or some other percent confident that the limits of the confidence interval do in fact surround the true but unknown population percent.

Keep in mind that a statistical fact obtained from a sample is a fact only insofar as that specific sample is concerned. As far as the corresponding population is concerned, the sample fact is only an estimate of an unknown, and usually more important, fact.

Now let us consider one of the most fascinating and most useful areas of statistical analysis, the analysis of *statistical relationships*.

13

Relationships: Causal and Casual

Statistics are no substitute for judgment.
—HENRY CLAY

One of the truly exciting goals of statistical analysis is that of discovering relationships among variables of interest. By the same token, evaluating relationships found by others is one of the most challenging tasks of the statistical critic because the measurement and interpretation of relationships is fraught with hazards, especially when assumptions about cause and effect are made. In this chapter we consider fallacious thinking about statistical relationships in general and about the assumption of cause and effect in particular. Before such fallacious thinking can be meaningfully discussed, however, I must acquaint you with some background concepts having their basis not only in statistics but, even more fundamentally, in philosophy.

The Murky Notion of Cause and Effect

Everyone thinks he knows what cause and effect is all about. Intuitively, the term seems crystal clear. Unfortunately for our sense of self-satisfaction, philosophers have an annoying tendency to demonstrate that many of the things we know we know we only think we know, as Lecomte du Noüy eloquently demonstrates in the following reflections on causation:

... let us take a cannon shot, for instance. Shall we say that the firing of a shell is "caused" by the small percussion cap, or by the movement of the soldier's hand which has pulled the string? Shall we say that the cause is the charge of powder? But without the movement of the hand this charge could have remained inert for centuries. Anyway, the movement of the hand can be replaced by a different mechanism and the explosion of the percussion cap could have been brought about by a very slight action, such as, for instance, the momentary interception of a feeble ray of light by the wing of a fly ... this slight ray of light ... will have played as important a part in the shooting as the charge of powder.

Neither can we say that the workers who manufactured the powder, or the chemical engineers who invented it, or the builders of the factory, or the capitalists who gave the funds to build it, or their parents or their grand-parents, etc., are responsible. And yet each one of them ... shoulders part of the responsibility, which gradually crumbles away, without every disappearing totally, and reaches back to the origin of the world.[1]

Vague Causes and Neat Labels

After the du Noüy quote, it might seem like an exercise in futility to try to place causes into nice neat categories, and, from a lofty philosophical standpoint, maybe it is. From a practical standpoint, however, the notion that if A happens, B will happen, or is more likely to happen, is often a most useful one. Maybe B is a crippling disease and A is an activity or condition somehow associated with the occurrence of that disease. If by eliminating A, we lessen the threat of this crippler, we have achieved something worthwhile even though we have not even begun to follow the full causal web in all its far-reaching and subtle ramifications.

From a practical standpoint, identifying different categories of causes is sometimes quite helpful in sharpening the thinking processes. Philosophers have identified the following four categories of causes:

Necessary Cause

A *Necessary Cause* is a condition that must be present if the event is to occur. An automobile must have fuel if it is to run. It won't go under its own power otherwise. In other words, if A is a Necessary Cause of B, then B will not occur without A.

Sufficient Cause

A *Sufficient Cause* is any condition that will bring about the event alone and by itself. If A is a Sufficient Cause of B, then B will always occur when

[1] *Human Destiny* (New York: Mentor Books, 1949), pp. 17–18. © Longman House and David McKay, Inc.

A occurs. If you cry every time you peel onions, then the act of peeling onions is a Sufficient Cause of crying. The act of peeling onions isn't a Necessary Cause of crying because other things can also lead to crying.

Necessary and Sufficient Cause

A *Necessary and Sufficient Cause* is any condition which will bring about the result and without which the event will not occur. That is, if A is a Necessary and Sufficient Cause of B, then B will occur if and only if A occurs. Maybe you never cry except when you peel onions. In such a case the act of peeling onions would be a Sufficient Cause of crying because it will do the job by itself, but it will also be a Necessary Cause as well, because nothing else moves you to cry. Needless to say, Necessary and Sufficient Causes are rare.

Contributory Cause

A *Contributory Cause* is a factor that helps to create the total set of conditions necessary or sufficient for an effect. Tension is thought to be a Contributory Cause of headache in that it contributes to a set of conditions which cause headache, but we can get a headache without having any recognizable tension. If A is a Contributory Cause of B, then B is more likely to occur when A occurs than when A does not occur.

Contributory Causes crop up more often in real-life situations concerned with the characteristics or activities of humans than do either clear-cut Necessary Causes or Sufficient Causes (and, of course, much more often than Necessary and Sufficient Causes) and are often quite difficult to track down and distinguish from other Contributory Causes. An error frequently made, and the source of a great many statistical fallacies, is to look for a simple, single cause of a phenomenon which is really the result of a combination of many Contributory Causes.

The Fallacy of Treating a Contributory Cause as if it Were the Sole Cause

As mentioned above, a frequent temptation is to single out a particular Contributory Cause and treat it as if it were the only cause—or, at least, the only one worth mentioning. This fallacy is really a form of jumping to conclusions, the subject treated in Chapter 10. Here is an example: A few years ago it was revealed that the life expectancy of the U.S. born male is a little less than 67 years, a figure which compares unfavorably with that of several other countries, especially Sweden, Japan, Czechoslovakia, and Israel. What follows is the explanation offered by four reasonably respectable sources as to why the American male shows such a penchant for dying early. What stands out in the following quotations is the considerable variety of plausible-sounding explanations based on single, seemingly all-important causes.

A magazine article stated: "The pressure to perform well in business looms ever larger as a reason why the life expectancy of males in the U.S. is only 66.7 years. . . ."[2]

Martin L. Gross pooh-poohs the pressure theory saying:

> Those who would glibly ascribe longevity failure of the American male to the pervasive stress of our society should first consider several factors: it has not been proved that life stress of any variety is a killer; no one has found a method of evaluating American affluence-stress with Israeli state-survival or Czechoslovak Communist-created stress; managing executives, who ostensibly are subjected to the apex of American stress, have a death rate some 20 percent *lower* than the union-protected plumbers.[3]

Gross himself emphasizes poor medical practice on the part of American doctors that produces a substantial gap between the present *potential* state and the *realized* state of the medical arts.[4]

Jean Mayer emphasizes neither stress nor poor medical craftsmanship. Instead, he attributes the relatively poor longevity of the American male principally to too much fat in his diet:

> Even without additives . . . the ordinary American's diet is dangerously unbalanced—rich in fat and poor in nutrients . . . the proportion of fat in the diet has swollen from 25 percent in 1900 to more than 40 percent today. In certain groups, such as college students and businessmen, it is well over 50 percent. The result has been massive heart disease and other weight and nutrition-related ailments, a life expectancy rate that dropped the American male from 11th place in 1949 to 37th place in 1966 and a spectacular wave of nutritionally unbalanced fad diets based on anything from the martini to the banana.[5]

Carlton Fredericks also places the blame on faulty diet, but, in his view, carbohydrates—especially sugar—not fats, are the chief culprits:

> Indeed, the arithmetic of diet would make it highly improbable that we significantly increased our fat intake in the last century. For during that century, on the basis of statistics which are above reproach, we have, instead,

---

[2] *Time*, July 18, 1969, p. 75.

[3] *The Doctors* (New York: Dell Publishing Co., Inc., 1967), p. 16. © Random House Inc., Alfred A. Knopf, Inc.

[4] *The Doctors*, pp. 15–16.

[5] *Life*, November 28, 1969, p. 44.

vastly increased our intake of *sugar*. The rate has climbed from twelve pounds per capita per year about a century ago, to *over one hundred pounds* per person per year in 1964. If we had simultaneously and proportionately increased our fat intake—which would almost be mandatory if indeed the fats are to be held responsible for the penchant of the American male to leave behind a rich widow—then obesity should not be the problem of a minority, but a characteristic of the entire nation.[6]

Undoubtedly, other plausible explanations have also been advanced.[7] I have presented only a few that I found with a bare minimum amount of research. All spokesmen agree that U.S. male longevity is not what it could be. But each emphasizes a single, seemingly all-important reason rather than acknowledging that there might be several more-or-less equally important Contributory Causes at work. In fairness to the sources quoted, I must mention that none is quite so lopsided in his views as the above quotations, lifted out of their respective contexts, might suggest. Still, the disparity of viewpoints used to explain the same fact and the convictions with which each explanation is offered are so marked that these quotations serve quite nicely to demonstrate that people, even experts, are strongly predisposed toward oversimplification. Beware of those who would oversimplify. Speculation about causes is healthy when used as a basis for painstaking hypothesis testing. But oversimplified speculation can also be harmfully misleading when presented as fact.

A similar kind of oversimplification, but in a grosser from, is an old standard of both advertising and politics. I trust you have seen before-and-after advertisements—you know, the kind with a blurry picture on the left showing a girl who must weigh at least 500 pounds and who also has fuzzy, unkempt hair, crooked teeth, and an unpleasant complexion. This is the "before" picture. The "after" picture, on the other hand, is bright and clear and shows a trim chick without an excess pound of flab and with faultless hair, teeth, and complexion. The advertising copy calls attention to the fact that the girls are really one and the same and that the "after" version owes her new-found beauty—and resulting happiness—to such-and-such reducing plan. Of course, the copy doesn't mention that the girl tried a variety of reducing aids in the process of becoming more streamlined. (The copy also fails to explain how this particular reducing plan straightened out the girl's teeth and made her hair more manageable or helped the clarity of the photography.) There is a kind of statistical fallacy analogous to such before-and-

---

[6] *Dr. Carlton Fredericks' Low-Carbohydrate Diet* (New York: Award Books, 1965), p. 37.

[7] For instance, one magazine article, though not making any international comparisons, reported that a study designed to determine factors which might contribute to deaths from strokes and heart attacks showed that too much sleep is responsible for many such deaths. (*Time*, October 25, 1968, p. 64.)

after advertising. This fallacy involves giving all the credit for a change to a single cause. Here are two examples from advertising:

A grocery chain pointed out that, although the Consumer Price Index for a specific area had increased sharply in the past month, the food-at-home component of the index had actually decreased. Then the ad asked: "Aren't you glad we went discount?"

Another advertisement asserted that many doctors have given up smoking without straining their will power through the use of a certain tablet that helps eliminate the need for nicotine and, hence, the desire to smoke. It went on to say that less than two percent of the 150,000 people who tried the product still smoke. The advertisement conveyed the impression that the tablet was solely responsible for the success of these people in giving up the habit when in reality at least some of these ex-smokers undoubtedly used a variety of methods.

In the context of politics this oversimplification fallacy usually takes the form of a politician's taking all the credit for something good or, conversely, giving his opponent all the blame for something bad. Governor Rockefeller of New York, for example, speaking of the adequacy with which his state fulfills certain key social responsibilities pointed out in an interview, ". . . the state this year is providing $500 million more for education than when I took office—in other words, a 90-percent increase in just four years."[8] Governor Rockefeller might indeed have had a great deal to do with this impressive increase. The subtle insinuation, however, that all of the increase had occurred because he was governor (and would not have occurred had he not been governor) is almost certainly an overstatement of the truth. At least some of

---

[8] *U.S. News and World Report*, April 8, 1968, p. 108.

the increase would undoubtedly have been made under a different governor
—and, who knows, maybe even a greater increase would have occurred!
The flip side of this strategy is to call attention to something that happened
during an opponent's term in office and speak as if the opponent had been
entirely to blame. Presidential aspirant and eventual winner, Richard Nixon,
in a campaign relatively free of mudslinging, used this tactic against then
Vice-President Hubert Humphrey in 1968:

> Hubert Humphrey defends the policies under which we have seen crime
> rising ten times as fast as the population. If you want your President to con-
> tinue a do-nothing policy toward crime, vote for Humphrey. Hubert Hum-
> phrey sat on his hands and watched the U.S. become a nation where 50% of
> the American women are frightened to walk the streets at night.[9]

Who could possibly be more surprised than a Vice-President, the butt of so
many jokes about ineffectual political positions, to learn that he had possessed the kind of power that Nixon was in effect attributing to Humphrey?

Some Thoughts on Tracking
Down a Cause

As an interpreter and critic of statistical information, you presumably
wouldn't be responsible for tracking down the cause(s) of something yourself. Nevertheless, the suggestions to follow might prove helpful when attempting to evaluate conclusions regarding causation that others make.

Suppose that a peculiar new disease breaks out on a university campus.
The disease so far has been largely confined to graduate students and the
main symptom is the occurrence of fits of uncontrollable laughter. Suppose
further that you have been given the assignment of finding the cause of this
bizarre afflction so that it can be brought under control. What would you do?

Undoubtedly, your first step would be to determine what characteristic or
activity the afflicted students have in common. However, if this is a large
university so that, even if only a small proportion of the students have the
symptom, it could still mean a rather large number of students, you could not
possibly do a thorough-going case study, complete in every possibly relevant
detail, for all the afflicted students. In order to cut some corners while still
keeping your goal clearly in mind, you do two things: First, you select a
random sample of students who show the peculiar symptom. Second, you set
up some broad hypotheses, based upon your knowledge and past experience,
to explain, tentatively, what might be causing the compulsive laughter. You
remember hearing stories about laughing gas, one of the early anesthetics,

[9] *Time*, November 1, 1968, p. 15.

and can't help noting certain similarities between the behavior of the people in these stories and the behavior of the afflicted graduate students. You might set up a broad hypothesis along lines of "The afflicted students are breathing something that is producing the uncontrollable laughter." You would use this tentative hypothesis to determine what characteristics or activities are relevant to your problem.

Assume that you find, reassuringly, that all graduate students in your sample have a characteristic in common—they are all taking classes from Professor R. E. Tort, an especially innovative chemistry teacher. Tort himself has also been a victim of occasional laughing spells, you discover. You are not yet dead certain that Professor Tort or his experiments are related to the affliction in a causal way; the students and Tort might have other things in common as well. Still, you have an idea worth pursuing. How might you go about testing your hypothesis? The great utilitarian philosopher, John Stuart Mill, offered some helpful suggestions.[10]

Testing by the Method of Agreement

According to Mill, if you could demonstrate that all cases have one and only one characteristic in common, then this characteristic is connected with, or is itself, the cause. Relating this principle to the problem at hand, you

[10] *Philosophy of Scientific Method* (New York: Hafner Publishing Company, 1950). Book III, Chapter VIII is the chapter most relevant to the present discussion.

would say to yourself, "If I can demonstrate that all those students with the affliction are enrolled in Professor Tort's chemistry classes and that they have nothing else in common, I can conclude that there is something about these particular chemistry classes that is related to the laughing sickness." Of course, you probably can do nothing of the kind since the phenomenon of interest has been taking place outside a controlled laboratory environment where other variables can be rigorously controlled. People will ordinarily have more than one characteristic in common. If, however, you can show that the afflicted students in your sample do have this one characteristic in common, even if it is not the only one, this fact should be sufficient to encourage you to keep looking in the direction of Tort's chemistry classes for an explanation. Of course, if some of the ill students are not taking chemistry and spend no time at all in close proximity to the chemistry class-rooms or lab, then this would lead to elimination of Tort's activities as a Necessary Cause.

Testing by the Method of Difference

You might also think in terms of a modification of another of Mill's principles, the method of difference. According to this method, a given characteristic is the cause or is connected with the cause of a phenomenon if it can be shown that (1) the specific characteristic is present in all cases where the phenomenon occurs; (2) the specific characteristic is absent from all cases in which the phenomenon does not occur; and (3) all other char-acteristics are the same not only for cases where the phenomenon occurs but for cases where it does not occur as well. To apply this method you must obtain a sample of students who have not been given to fits of compulsive laughter. If you can show that the afflicted students and those not so afflicted differ only in that the afflicted ones are taking classes from Tort and the well ones are not and the two groups differ in no other respect, you will have applied the method of difference. But, again, this is a well-nigh impossible trick to pull off outside a carefully controlled environment. Still, the use of a sample of well students might be extremely helpful in eventually pinpointing the cause. But suppose that some of the well students are taking chemistry from Tort? In such a case Tort's classes would have to be eliminated as a Sufficient Cause, but they still might constitute a Necessary Cause.

Testing by Concomitant Variations

This method of pinpointing causation is the one most relevant to cor-relation analysis, the branch of statistics concerned with measuring the strength of the relationship between two or more variables. According to the method of concomitant variations, when, within limits, two phenomena vary together, one phenomenon is the cause of the other one or is in some way connected with the cause. Hence, if students spending several hours a day in the chemistry lab working on experiments related to Tort's classes have more frequent and/or more severe laughing spells than those who spend little time each day in this pursuit, you would be led to suspect that there is something about Tort's assignments that produces this bizarre symptom. Actually, this method, like the others discussed above, works much better under carefully controlled conditions than it does in the real world where a great many variables may be playing upon the phenomenon under study. In the real world there are just too many ways that things can vary together without necessarily being causally related in any direct sense. Another way of stating the same point is that in the real world, the risk of committing the *post hoc* fallacy is great. I will return to this assertion and discuss the meaning of the *post hoc* fallacy after a brief but necessary digression on some fundamentals of correlation and regression analysis.

A "Cook's Tour" Around Correlation and Regression Analysis

On the off-chance that you have already jumped to the hasty conclusion that you are going to learn a great deal about correlation and regression analysis from reading this section, let me disabuse you at the outset. Formal correlation and regression analysis, although fascinating, or so some of us believe, is a multifarious, highly technical subject. Entire books much larger than this one have been written about this branch of statistical analysis.[11] My aim in writing this section is simply to acquaint you generally with a few basic tools and definitions that just about have to be brought into any discussion of correlation and regression fallacies. My efforts in this direction will be facilitated through the use of a simple, hypothetical example.

Let us suppose that you believe there is a relationship between the market prices of common stocks and the per-share earnings of the related companies. You wish to get confirmation for this belief, if possible, and, at the same time, develop a mathematical equation which describes the relationship. Your thinking is that, if you can make a reasonably accurate estimate of a company's earnings, you should be able to substitute the earnings figure into the mathematical equation and crank out an estimate of the average market price of the company's common stock and, in turn, judge the stock's investment potential. Of course, in order to do this you must develop the equation using data for a past period. Let us say, for simplicity's sake, that you pick

[11] One very readable book is Mordecai Ezekial and Karl A. Fox, *Correlation and Regression Analysis*, 3rd ed. (New York: John Wiley & Sons, Inc., 1959.)

a random sample of ten companies from among all those listed on the New York Stock Exchange and get the earnings figures for each company for the previous year. Earnings data are used as the independent or X variable. You also obtain average market prices for the same ten stocks for the previous year. Market prices are used as the dependent or Y variable. Assume, unrealistically, to be sure, that the data are as shown in Table 8.

Table 8. Earnings Per Share and Average Market Price of Ten Randomly Selected Stocks for the Year 19_____

	Average Market Price (Y)	Earnings Per Share (X)
	60	2
	70	3
	50	2
	70	2
	30	1
	30	1
	40	1
	70	2
	80	3
	90	3
Total	590	20

Source: Hypothetical data.

The Scatter Diagram

Does a relationship exist at all? If so, is it linear or curvilinear? Is it direct (positive) or inverse (negative)? Is it quite strong or rather weak? These are four questions that one usually seeks to answer when undertaking formal correlation and regression analysis. Pretty good tentative answers can be obtained through the use of an enormously helpful and refreshingly simple device called the *scatter diagram*.

A scatter diagram for our data is shown in Figure 13–1. As you can see, the scatter diagram is made up of several dots—one dot for each pair of numbers in Table 8. Just by eyeballing the scatter diagram we can learn a great deal. For example, the fact that the dots form a pattern rising toward the right indicates that as X (earnings) increases, Y (market price) has a tendency to increase also. Such a pattern of dots tells us that (1) a relationship does indeed exist, and (2) it is direct or positive (in contrast to inverse or negative, a situation that would be revealed by a pattern of dots declining toward the

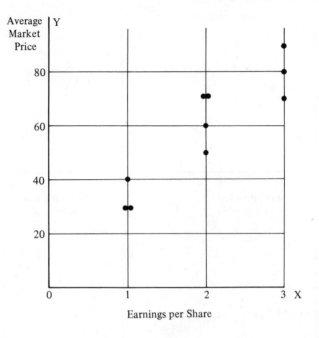

Figure 13–1

right). We also see from the scatter diagram that the relationship might be described adequately by a linear rather than by some kind of curvilinear function. That is, if a line were drawn through the center of the dots, it would probably be straight rather than curved. A fairly strong relationship can be deduced from the relatively small amount of scatter around an imaginary line through the center of the dots. Just how strong the relationship is can be measured precisely, as I'll soon demonstrate. Figure 13–2 displays a few other possible patterns that are sometimes obtained when scatter diagrams are constructed.

Fitting a Straight Line to the Data

Regression analysis is concerned with the form of the relationship between the variables. That is, we have already noted that a line through the center of the dots would probably be straight. If you wished to express the form of the relationship more precisely, you could do so by obtaining a mathematical equation of the form $Y_c = a + bX$, where Y_c stands for computed Y as distinguished from actual Y, a is the value of Y_c at the point where the straight line cuts the vertical axis, and b is the slope of the line. Such an equation was developed by me for the data in Table 8. The equation is

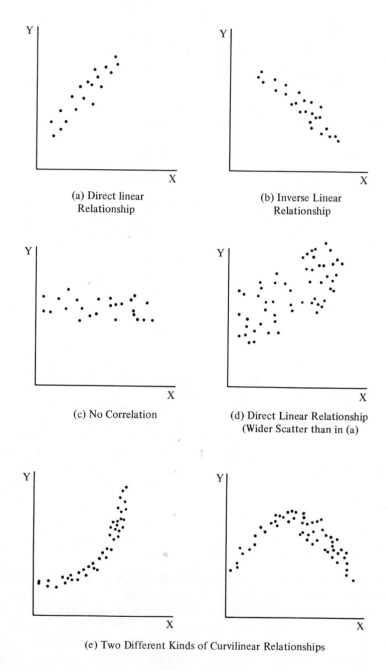

(a) Direct linear
Relationship

(b) Inverse Linear
Relationship

(c) No Correlation

(d) Direct Linear Relationship
(Wider Scatter than in (a)

(e) Two Different Kinds of Curvilinear Relationships

Figure 13–2

$Y_c = 12.33 + 23.33X.$[12] The figure 12.33 is read \$12.33 and serves to place the line at the right height for it to pass through the dots. The figure 23.33 indicates that as earnings per share increases by \$1.00, market price increases, on the average, by \$23.33. The straight line fitted to the dots is shown in Figure 13–3.

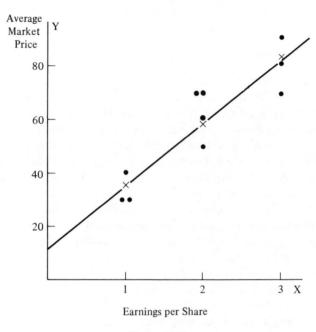

Figure 13–3

Measuring the Strength of the Relationship

How strong the relationship is—that is, how dependably the two variables vary together—is what correlation analysis as distinguished from regression analysis attempts to measure. The strength of the relationship can be measured by the so-called *coefficient of correlation*, often denoted by r. The coefficient of correlation may have any value between -1 (meaning a perfect

[12] If you happen to be a formula buff, you might enjoy these, the ones used to obtain the a and b constants:

$$b = \frac{\Sigma\, XY - (\Sigma\, X)(\Sigma\, Y)/n}{\Sigma\, X^2 - (\Sigma\, X)^2/n}$$

$$a = \frac{\Sigma\, Y}{n} - b\frac{\Sigma\, X}{n},$$

where Σ means "summation," Y is the dependent variable, X is the independent variable, and n is the sample size. These formulas provide a and b values such that the resulting straight line minimizes the sum of the squared deviations between Y and Y_c values.

inverse relationship) to $+1$ (meaning a perfect direct relationship). A value of zero means that there is no relationship at all. Clearly, a coefficient of correlation close to 1.0 (or -1.0 in the case of an inverse relationship) is more indicative of a dependable relationship than is a coefficient of correlation close to zero. For the example under discussion, the coefficient of correlation is slightly over 0.9, a condition suggesting a rather strong relationship.[13]

The Matter of Significance

So far we have focused on data for a sample of ten common stocks. If such data were your ultimate concern, the values of a, b, and r could be depended upon as perfectly good descriptive measures. However, since we have been assuming all along that you took pains to select a truly random sample of ten stocks from among all those listed on the New York Stock Exchange, it must be the entire list—i.e., the entire population—of stocks traded on this major exchange that is really your ultimate interest. If so, you must realize that the a, b, and r values are only estimates of the true but unknown population counterparts of these measures. It is highly unlikely that the sample measures already obtained and the population measures, if known, would coincide exactly. Some implications of this fact can probably be discussed most meaningfully by focusing on r, the sample coefficient of correlation. As mentioned above, a value in excess of 0.9 is suggestive of a strong relationship. But before you can place much confidence in it, you must make certain that it does reflect a valid statistical relationship for the total population of stocks rather than being merely a sampling fluke. Because your sample of ten stocks was selected at random, it is possible that earnings and market price just happen to correlate strongly for the ten specific stocks selected even though any such relationship is completely absent in the population. Figure 13-4 illustrates the issue. The population values are seen to be uncorrelated, but the sample figures show a strong correlation simply because random sampling just happened to provide you with misleading data. To make certain—or, at least, pretty certain—that the population values are not devoid of correlation, we apply a so-called test of significance to r. Such a test of significance tells us the chances of getting a sample coefficient of correlation as high as 0.9 by sheer chance even though the coefficient of correlation for the entire population is a flat zero.

A test of significance was applied to our 0.9 figure, and significance was found at the .001 level. What this means is that we would get a coefficient of

---

[13] The formula used to compute r is

$$\frac{\Sigma\, XY - (\Sigma\, X)(\Sigma\, Y)/n}{\sqrt{[\Sigma\, X^2 - (\Sigma\, X)^2/n][\Sigma\, Y^2 - (\Sigma\, Y)^2/n]}}$$

where Σ means "summation," Y is the dependent variable, X is the independent variable, and n is the sample size.

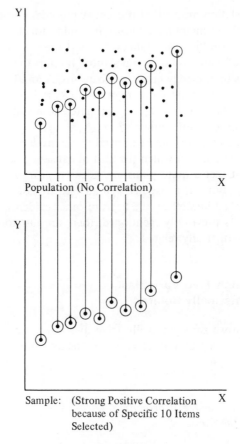

Figure 13–4

correlation as high as 0.9 or higher less than one time in 1000 random samples of size ten obtained from this same population. We still don't know for certain that our relatively high coefficient of correlation was not obtained simply as a sampling fluke. But since the odds are strongly against this possibility, we accept the *r* value obtained from the sample as indicating a valid relationship.[14]

The Post Hoc Fallacy

The term *"post hoc* fallacy" is a shortened form of the Latin expression *post hoc ergo propter hoc*, which means "After this, therefore because of this."

---

[14] The population, however, does not necessarily have a coefficient of correlation as high as the 0.9 found in the sample. All we establish using such a test of significance is that it is highly unlikely that the population coefficient of correlation is zero.

This literal translation means that if event A precedes event B, then A must be the cause of B. ("I might have known it would rain; I washed my car this morning.") The term "*post hoc* fallacy" is used today to mean something a little more general than this literal translation. It refers to the error one makes when he assumes that, merely because two events, A and B, occur together or in succession—or, in terms more relevant to correlation analysis, merely because two variables, *X* and *Y*, are correlated to a statistically significant degree—one is *cause* and the other *effect*.

The *post hoc* assumption can be applied to a considerable variety of situations ranging from those in which the idea of causation seems outlandish to those in which it seems quite plausible. Clearly, *post hoc* reasoning is not inevitably wrong. But one is usually well-advised to assume that it is wrong until he has satisfied himself on the basis of other evidence that causation is indeed present. As previously mentioned, there are just too many ways for variables to be statistically related.

Ways in Which Two Variables Can Be Statistically Related

In this section we survey and illustrate five ways that two variables can correlate strongly, only two of which are perfectly compatible with the *post hoc* assumption.

Two Variables, X and Y, Correlate
Because X Is the Cause of Y

The *X* variable may actually be the cause or be intimately associated with the cause of *Y* as *post hoc* reasoning assumes. For example, there is known to be a strong correlation between temperature, measured in degrees Fahrenheit, and the frequency with which crickets chirp. According to Croxton and Cowden,[15] if you were to multiply the degrees Fahrenheit by 3.78 and subtract 137 from the result, you could estimate the number of chirps to be expected from a cricket in one minute with considerable accuracy. (That is, $Y_c = -137.22 + 3.777X$, where Y_c is the computed or estimated value of number of chirps in a minute and *X* is the temperature in degrees Fahrenheit.) That the two variables are strongly correlated is attested to by the fact that the coefficient of correlation is an impressive 0.9919, where, you will recall, 1.0 is the highest possible value this measure can have.

It seems infinitely more sensible to suppose that variation in the tempera-

[15] Frederick E. Croxton and Dudley J. Cowden, *Applied General Statistics*, Second Edition (Englewood Cliffs, N. J.: Prentice-Hall, Inc., 1955), pp. 451–54.

ture somehow produces changes in the number of times a cricket chirps in a minute than to assume the reverse. We conjure up a ludicrous picture indeed when we try to imagine a cricket increasing the speed of his chirping, thereby driving up the temperature. (If he were to stop chirping altogether would his reticence usher in a new Ice Age?).[16]

Two Variables, X and Y, Correlate Because Y Is the Cause of X

Sometimes the presumed cause is really the effect and the presumed effect really the cause. I am reminded of Chanticler, the rooster hero of the Edmond Rostand play by the same name. Early in the play, the proud cock suffers from the delusion that his song directly evokes the day. Without his crowing, he believes, the sun would not rise and the day could not begin. Fortunately, in a cause-and-effect sense, if in no other sense, the tale has a happy ending. Chanticler learns that he really has nothing to do with summoning the new day but decides to keep on crowing anyway to proclaim its arrival.

Imagine yourself employed by a company as a marketing researcher. Management wishes to know whether the firm's sales could be increased if it were to spend more money on advertising than it customarily has. Since your assignment is to tell management how things might be under (as of right now) hypothetical conditions, you decide to determine what the experience of other firms has been. You select a random sample of businesses and run a correlation analysis on sales in dollars and advertising expenditures using the latter as the independent variable and presumed cause. You get results as shown in Figure 13–5.

Let us suppose you discover a strong positive correlation between the two variables. What do you conclude? Should the company be spending more money on advertising or not? I have intentionally been leading you down the primrose path—maybe. Actually, you wouldn't really know from such evidence whether more advertising was warranted because you couldn't be completely certain that you had assigned the cause-and-effect roles properly. Perhaps it is the businesses with the highest volume of sales that are the ones most inclined to spend lavishly on advertising. In other words, there might be a causal relationship, but the chain of cause and effect might be running from sales to advertising rather than from advertising to sales.

Along this very line, a magazine article stated: "One study of 105 brands showed that the five brands that averaged 13 pages of advertising a year gained 6.8% in purchases, whereas the 17 that average five pages a year

---

[16] Actually, I am being facetious. For reasons known only to crickets, all chirping ceases at temperatures below about 45 degrees.

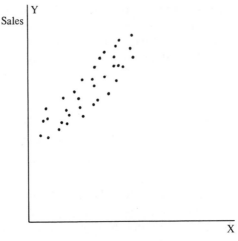

Advertising Expenditures

Figure 13–5

gained 5.5%, and the 40 that did no advertising in the magazine market declined 6% a year."[17] But did healthier sales result from greater amounts of advertising or did healthier sales encourage greater amounts of advertising? We know too little, of course, to draw a firm conclusion. The point is: The chain of cause and effect might, in any given instance, be running in just the opposite direction than we are being encouraged to believe.

A news magazine reported: "When a family gets a second or third car, you might expect that miles of driving per car would be reduced—but it doesn't work out that way. A study by economists of the Bureau of Public Roads indicates that families with only one car average 9,900 miles in a year but those with two or more average 10,000 miles for each car."[18] The wording seems to suggest that the more cars a family owns, the more driving they are inspired to do. It might make more sense—or, at least, just as much—to interpret these figures as indicating that families which do a great amount of driving feel a need for more cars.

Two Variables, X and Y, Correlate Because they Interact with Each Other

Returning to our marketing research study of the relationship between advertising expenditures and sales levels, a more realistic interpretation than one dependent upon a simple one-directional cause-and-effect relationship

[17] Roger Barton, "Second Thoughts," *Media/Scope*, January 1968, p. 72.

[18] *U.S. News and World Report*, March 4, 1968, p. 86.

might be that advertising and sales interact in such a way that X is sometimes cause and sometimes effect, and the same with Y. That is, maybe advertising helps to bolster sales of small firms which, as their sales and, hence, size increases, can afford to spend more liberally on advertising. The additional advertising helps to boost sales even more and, in turn, still more lavish spending on advertising can be afforded. And so it goes. You, of course, would not be aware of this spiraling effect if your research were limited to the simple correlation analysis assumed above. (The scatter diagram in Figure 13–5, for example, might reflect such interaction, but you can't tell by looking at it.) Nevertheless, the point is valid enough: There could be cause and effect present, but it might run from X to Y through a range and from Y to X through a range.

Supply-and-demand relationships suggest another way that interaction can be present. A high demand for a product will lead to higher prices and, in turn, an increased supply. But as supply increases, downward pressure is exerted on prices. Lower prices increase demand, and the cycle begins afresh.

Two Variables, X and Y, Correlate by Chance

Does it sometimes seem to you that everytime you get comfortably submerged in the bathtub the telephone rings? Of course, if you take a great many baths and get a great many calls, you might reasonably expect such a condition to exist. But someone like myself who takes relatively few baths and receives relatively few calls might be justified in thinking that such a coincidence—shall we call it the Dagwood Bumstead syndrome?—should occur less often than seems to be the case. I use the word "coincidence" as that is almost certainly what it is. I feel reasonably sure that at my house, running the bath water doesn't set up some kind of electrical charge which travels through the air and causes the telephone to ring. I feel equally certain that someone's intention to phone me in a few minutes is not picked up by me through some sort of extrasensory-perception mechanism and translated into an undeniable compulsion to take a bath. At best, any cause-and-effect system present would work through the fact that people who know me and who might have reason to call me at home know that I tend to be at home during certain hours of the day and away from home during certain other hours. They will naturally be predisposed to call at a time when they expect me to be at home. Since I take 100 percent of my baths at home, it perhaps isn't too surprising that a bath and a telephone call might occur at about the same time.

In reality, I probably don't get all that many calls while taking a bath. When it does happen, I find it annoying and therefore tend to remember the event. When it doesn't happen, I don't even think about it—then, or later. But enough of this. The point is that sometimes things do happen by coincidence. If the number of observations is small, a few such coincidences may convey the impression of a valid relationship. Usually, however, the relationship vanishes when more observations are obtained.

Putting the preceding point in statistical terms: *Whenever correlation analysis is applied to sample data, as is often the case in practice, the risk exists that sample data will correlate even though there is no real correlation present in the corresponding population(s).* This risk is especially great when the sample size is small. Tests of significance which statisticians employ when the sample has been selected at random (one of which was discussed briefly in a previous section of this chapter) serve to flush out many "chance" relationships. However, even they cannot put the lie to all of these because the foundation of all such tests is probability theory. Whenever one depends on probability theory, he must be aware that once in awhile even highly unlikely events will occur by chance.

Two Variables, X and Y, Correlate Because
they Are Both Effects of a
Third Variable Outside the Analysis

The story is told about a man who wrote a letter to an airline requesting that their pilots cease turning on the little light that says FASTEN SEAT BELTS because every time they do, the ride gets bumpy. Needless to say, leaving the light off would not make the ride any smoother. Both the light and the bumpiness are the result of a third variable apparently not taken into account by this letter writer, namely the existence of very real turbulence.

Before storks became candidates for extinction in recent years, there was reportedly a positive correlation between the number of stork's nests and the number of human births in northwestern Europe. This correlation rather than adding credence to the old dodge about how babies come into the world was simply a reflection of population growth. As population and, hence, the number of buildings increased, the number of places for storks to nest increased.[19]

One of my former statistics professors used to tell the story about a marketing research study conducted some years ago with a view to determining some of the causal factors underlying typewriter sales. The researchers discovered a strong positive correlation between number of typewriters sold and number of subscribers to a certain popular magazine in several geographical areas. The tentative explanation that "If you can't read the magazine, you probably can't use a typewriter" seemed a little strained. In reality, both typewriter sales and number of subscribers to the magazine were reflections of population size.

Croxton and Cowden relate the tale of a southern meteorologist who discovered that the fall price of corn is inversely related to the severity of hay fever cases.[20] This finding does not suggest, however, that the low price of corn causes hay fever to be severe nor that severe cases of hay fever bring about a drop in the price of corn. The price of corn is usually low when the corn crop has been large. When the weather conditions have been favorable for a bumper crop, they have also been favorable for a bumper crop of ragweed. Thus the fall price of corn and the suffering of hay fever victims can each be traced to the weather, but are not directly dependent upon each other.

Two Common Regression Fallacies

Because erroneous conclusions about causation are arrived at so depressingly often from observed statistical relationships, I have emphasized correlation analysis in this chapter. But fallacious conclusions from regression analysis, the branch of the subject concerned with describing the form of the relationship, are by no means rare. Therefore, in this section we shall consider two common regression fallacies: (1) extrapolation beyond the range of the observed data, and (2) *the* regression fallacy.

Extrapolation Beyond the Range of the Observed Data

"Your husband could dance with you for 11 minutes on the energy he'd get from 49 raisins. Think what would happen if he never stopped eating them."[21] Believe it or not, this advertisement appeared in at least one of the major ladies' magazines a few years ago. I can think of a lot of things that might happen to a man if he never stopped eating raising—and they are all very painful. Presumably, however, we are being invited not to contemplate the physical discomforts that would result from the uninterrupted eating of raisins but rather to imagine the boundless energy that would accrue. In other words, we are being encouraged to plug an X value of infinity into some equation not specifically stated in the ad and come to the conclusion that the resulting energy would also be infinite. (Anyway, that's how I interpret this strange ad. Perhaps you interpret it quite differently.) Accepting any such invitation is always potentially dangerous as can be demonstrated using a known regression equation.

When we discussed the example of the relationship between temperature and frequency of cricket chirps, we noted that this relationship can be described by the linear function $Y_c = -137.22 + 3.777X$, where Y_c is the

[20] Frederick E. Croxton and Dudley J. Cowden, *Applied General Statistics*, Second Edition (Englewood Cliffs, N. J.: Prentice-Hall, Inc., 1955), pp. 9–10.

[21] From California Raisin Advisory Board advertisement.

estimated number of chirps made by a cricket in one minute and X is the temperature in degrees Fahrenheit. The problems that can result from extrapolating the line beyond the range of the observed data can be illustrated most simply by focusing on the a value of -137.22. Does this figure mean that if the temperature is zero degrees Fahrenheit, a representative cricket will chirp minus 137 times in a minute? Obviously not. A cricket cannot chirp fewer than zero times in a minute and the temperature at which he shuts up completely happens to be around 45 degrees. One should be cautious when interpreting the a value as if it were the value that Y_c would have when X equals zero. This constant merely serves to place the height of the line appropriately so that it will go through the center of the dots. Any other interpretation is potentially misleading.

What about extrapolation in the other direction? That's risky too. Suppose, for example, that we wanted to know how many times a cricket would chirp in a minute when the temperature is 500 degrees Fahrenheit. Plugging the figure of 500 into our equation, we get Y_c equal to 1751.22 times. Is this a realistic figure? Again, obviously not. To begin with no one has made observations at temperatures in excess of about 75 degrees. So what happens at higher temperatures is not known. It seems reasonable to believe, however, that the number of chirps would be limited by the time constraint long before a temperature of 500 degrees was approached. That is, there is probably some absolute physical limit to the number of chirps that can be emitted in a minute. Not only that but, at some temperature lower than 500 degrees, the cricket would be roasted to death and, hence, do no chirping at all.

It is always risky to extrapolate the line of regression beyond the range of the observed data. To be sure, on occasion such extrapolation is necessary, but one should always be mindful of the high risk he is running of getting misleading results.

The Regression Fallacy

The preceding is *a* regression fallacy. What follows is a discussion of *the* regression fallacy. This fallacy can be traced to the very beginning of correlation and regression analysis. Sir Francis Galton during the latter part of the nineteenth century developed the rudimentary beginnings of this branch of statistical methodology. He also has the dubious distinction of committing the first regression fallacy having its basis in formal analysis. In his studies of hereditary traits, Galton pointed out the apparent regression toward the mean in the prediction of natural characteristics. He found, for example, that unusually tall men tend to have sons shorter than themselves and unusually short men tend to have sons taller than themselves. This fact suggested to Galton a "regression toward mediocrity" in heights from generation to generation. It was this reasoning, by the way, that prompted the term regression analysis in use today.

The phenomenon that Galton observed is real enough, but the assumption of "regression toward mediocrity" is faulty and has been appropriately termed the regression fallacy. A sounder explanation is that extreme values for a characteristic occur in part by chance; hence, the odds are that the genetic factors producing the unusual height or lack of height will not be passed onto the offspring. A convenient way of schematizing the underlying concept is to view tall people as belonging to four possible categories: (1) typical individuals from tall parents, (2) unusually short individuals from extremely tall parents, (3) unusually tall individuals from medium-sized parents, and (4) unusually tall individuals from extremely short parents. Group 3 is more numerous than either group 2 or 4, simply because medium-sized parents are more common than extremely tall or extremely short parents. Therefore, if we were to select a group of tall people and measure the heights of their parents, we would expect to find the parents shorter, on the average, than their offspring (though taller, on the average, than the general population).

To repeat a point made above: The regression phenomenon is very real and, as such, does not constitute a statistical fallacy. It is the assumption of a tendency toward mediocrity or, to put it another way, toward a reduction in dispersion in the total population, which is the fallacy.

One sometimes hears that extremely intelligent and extremely dense people tend to have children closer to the average I.Q. than the parents. Be that as it may, to conclude that the total population is, therefore, tending over the generations toward mediocrity is to commit the regression fallacy.

Frequently the regression phenomenon occurs when extreme figures for a given year are compared with related figures in a later year. For example, in any given year some business firms will experience unusually high profits relative to those of other firms because of nonrecurring factors such as extraordinarily high but short-lived demand for a particular product. If such firms do not fare so well relative to other firms in the following year, it should come as no great surprise. But neither should it be interpreted as a symptom of "regression toward mediocrity" in profits.[22] When attention is focused on firms at the extremes with respect to profits in one year, the firms under scrutiny are of two kinds: (1) those that are generally at the extremes and which can be expected to stay there, and (2) those that are generally at the center of the distribution but happen to be at the extremes for unusual short-term reasons and which can be expected to move back toward the center. The latter firms move the average toward the center, but their places at the

[22] This very form of the regression fallacy served as the basis for a strategic part of a book by Horace Secrist called *The Triumph of Mediocrity In Business.*

extremes tend to be taken by other firms that are normally near the center but which are influenced by short-term factors pushing them temporarily toward the extremes.

So much for fallacious thinking about statistical relationships. Let us now move on to consider briefly several other kinds of statistical fallacies which, although important, do not lend themselves conveniently to inclusion in any previous chapters.

14

Leftovers

Wisdom came to earth and could find no dwelling place.
— ENOCH

There are still several kinds of statistical fallacies that simply don't fit nicely under headings used in previous chapters but are sufficiently common that they merit treatment somewhere. Hence, this chapter on the Leftovers.

Faulty Deduction

I dealt briefly with deduction as a logical exercise in Chapter 12, but deduction as a way of reasoning statistically also deserves some attention. Often erroneous conclusions are reached because valid figures for a large group are applied to subgroups or samples for which they may be inappropriate. It is often stated, for example, that exports are unimportant to the United States economy because they invariably amount to less than 10 percent of the gross national product. Whatever merit this argument might have overall does not hold for some individual industries which depend mightily on exports.

Spurr and Bonini cite the case of an electric institute report stating that "industry's generating capacity in December was 5.1 percent above electricity demands." The statement was true enough for the total nation. However, some regions such as the Far West, which had been growing rapidly, were at that time short of power.[1]

A company asserted in its advertising that one out of the next 50 drivers coming your way is drunk. One-in-50 might be the correct proportion for the total population of drivers but not necessarily for the next 50 cars coming your way. In a report on alcoholism in business the following statement appeared: "One industrial psychologist says that he tells supervisors: 'If you've got 14 people working for you, at least one of them has a drinking problem. If you don't believe it then you're kidding yourself.' "[2] Again, the proportion cited might hold for the total population, but it doesn't necessarily hold for a specific sample of 14 employees.

A former student told me of an interesting off-beat example of faulty statistical deduction. The story concerns a pot-luck dinner held in the church he attends. Families with last names beginning with A–B were asked to bring rolls or bread and butter; families with last names beginning with C–J were asked to bring a salad; families with last names beginning with K–O, dessert; and P–Z, a main dish of meat or vegetables. These assignments were made on the basis of the distribution of names on this church's rolls. If every family had shown up, or even a representative sample of families, the meal would have been well-balanced. As it turned out, this student informed me, the sample of those in attendance was anything but representative and, as a result, the dinner consisted of a long series of desserts.

Omission of Strategic Details

The story is told about an event during a trial following a calamitous wreck at a railway crossing. A trainman testified that he had signaled vigorously by waving a lantern. His animated demonstration of the vigor with which he waved the lantern greatly impressed the jury which eventually found the railway and its personnel innocent of negligence. After the trial, the lawyer for the railway commended the man for the effectiveness of his testimony. The trainman, obviously greatly relieved, said, "Thanks. But for a while there I was afraid the other lawyer was going to ask whether the lantern was lighted."

As the courts have long recognized, it is not enough that the truth and nothing but the truth be presented; it must also be the whole truth. Often statistical evidence tells the truth and nothing but the truth but is still faulty because the whole truth has been withheld. Not infrequently it is the facts

[1] William A. Spurr and Charles P. Bonini, *Statistical Analysis for Business Decisions* (Homewood, Illinois: Richard P. Irwin, Inc., 1967), p. 10.

[2] *Business Week*, October 26, 1968, p. 97.

withheld that are more important to the shaping of a valid conclusion than the facts presented. Here are a few illustrative examples:

The following advertisement appeared in the *Public Relations Journal*:[3]

IN 4 OUT OF 5 CORPORATIONS TOP PUBLIC RELATIONS MEN REPORT DIRECTLY TO THE PRESIDENTS OR CHAIRMEN OF THE BOARD

According to a survey among corporate chairmen and presidents made by_____, Director, The American University Center for the Study of Private Enterprise, public relations men have the ear of top management. Here are the facts:

To Whom Does PR Man Report	All Companies (%)	Consumer-Oriented Companies (%)	Non-Consumer-Oriented Companies (%)
(1) President/Chairman	78	78	79
(2) Senior Vice President	4	6	1
(3) Vice President—Marketing	3	2	4
(4) Executive or Group Vice President	4	5	4
(5) Vice President—Finance	2	2	1
(6) Vice President—Advertising	1	1	0
(7) Assistant to President	1	0	1
(8) Manager, Corporate Personnel	1	0	1
(9) Vice President—Secretary	1	0	1
(10) Vice President—Operations	1	0	1
(11) Did Not Respond	6	7	4

What this advertisement fails to mention is that many small companies don't have all the positions listed. The majority of corporations in the United States have only four or fewer officers and, of these, the president is often the only one with direct responsibility for the company's image. Granted, the ad is probably telling the truth. But it is not telling the whole truth because the listing of ten separate positions encourages the reader to envision very large businesses and, in turn, get the impression that the public relations men in giant corporations have a direct line to the president or chairman of the board—an exaggeration of the truth at the very best.

A national magazine reported that a group of Colorado school teachers had been given a test in history and had failed with an average grade of 67. This result was interpreted as an indication that Colorado teachers in

---

[3] September 1968, p. 28.

general were deficient in history. However, an official of the Colorado Education Association clarified the matter. It seems that only four teachers had been given the test. Three had made the respectable average score of 83 and the fourth only 20, bringing the average of the four down to 67.[4]

Wallis and Roberts cite the example of an unusual deluge of rain in Palo Alto, California, on July 25, 1946. The incredible facts are as follows: Nineteen times as much rain fell between 6 A.M. and noon as during all the preceding Julies since the weather station opened in 1910. Another way of stating the situation is: In six hours, 19 times as much rain fell as in a 26,784-hour period, a rate during those six hours about 85,000 times "normal." The facts presented in this fashion makes the experience sound like one of this nation's all-time great downpours. But, as Wallis and Roberts point out, the "deluge" consisted of only 0.19 inches; the only measurable rain in all the 36 previous Julies was on one occasion when 0.01 inches fell.[5]

Another vintage example has to do with data presented to a congressional investigating committee in 1933 by proponents of a higher tariff. Fearing that the devaluation of the British pound sterling in 1931 might be giving the

[4] Cited in William A. Spurr and Charles P. Bonini, *Statistical Analysis for Business Decisions* (Homewood, Illinois: Richard P. Irwin, Inc., 1967), p. 9.

[5] W. Allen Wallis and Harry V. Roberts, *Statistics: A New Approach* (Glencoe, Illinois: The Free Press, 1956), p. 83.

members of the sterling block an undue advantage in foreign trade, they argued that American producers should be protected by a higher tariff. They supported this stand by producing figures showing an increase of 2500 percent in imports of canned salmon into the United States within a few months after the devaluation of the pound. Pig iron imports from the United Kingdom had risen 611 percent and some cotton imports, 1238 percent.[6]

What the proponents failed to mention, however, was that the increase in canned salmon imports was only 44,000 pounds, a negligible amount when compared with 300,000,000 pounds then being produced by American packers. The 611 percent increase in pig iron represented only 400 tons valued at about $4500; and the dollar figure for cotton imports was a mere $982.

"Statistics" and Other Magic Words

Related to the fallacy of omitting strategic facts is the fallacy of omitting sources. Purveyors of faulty statistics are well aware that a peculiar but, alas, common characteristic of the statistically unsophisticated is to be overawed by statistical analysis to the point that they will accept practically any argument preceded by "statistics prove," "according to statistics," "recent statistical studies show," "according to a nationwide poll," "recent research indicates," and so forth. Needless to say, the statistical critic should know the source of the "evidence" cited and evaluate the source's credentials if at all possible. If all one has to say to convince an audience of something is that "Statistics prove blah, blah, blah," then the possible abuses are limitless.

Uncritical Acceptance of Computer Results

So useful has the computer become in all branches of statistical analysis that there may be some tendency to forget that even it has its limitations. The computer cannot work magic—not yet anyway. It will do only what it is instructed to do, and the validity of the results is determined by the accuracy and adequacy of the data put in and the wisdom of the people writing the instructions. Granted, the computer can perform a great many calculations much more rapidly than mere mortals can do them. Nevertheless, speed of computational work is not the same thing as infallibility in aiding with the decision-making process. A statistical critic, of all people, should guard against being overawed by the news that certain information was turned out by a computer. The mere fact that computers are being used these days even to cast horoscopes should be ample proof that a computer is no more immune to spewing out nonsense than are real flesh-and-blood people.

[6] *The New York Times*, February 22, 1933.

Superfluous Data

Sometimes too much is said. That is, information is volunteered that isn't directly relevant to the point being made but is stuck in because it looks or sounds impressive. Here is an example that seems to be of this ilk:

The total cost of transportation for the nation is approximately 20 percent of the GNP or about $150 billion annually. Approximately half of the expense for freight transportation may go toward paying for what one member of the transportation industry calls the "grand tiddly-winks game of shuffling goods on docks, platforms, between vehicles, and in other side expenses like packaging, damage claims, insurance and the like.[7]

Since the author doesn't tell us what proportion of the impressive-appearing $150 billion is accounted for by freight transportation, what have we gained from having been presented with this large figure in the first place? The author has encouraged us to think in terms of astronomical costs and then tells us about a situation related to only some unknown fraction of those costs. However valid the point, the author himself appears to be engaging in some statistical "tiddly-winks."

Crazy Ratios

Sometimes ratios which are perfectly accurate from an arithmetic standpoint lead to bizarre impressions and are, therefore, worthy of being included among the Leftovers. Someone might say to you, for example, "Since about $3\frac{1}{2}$ million babies were born in this country last year and since the population was about 210 million or so, we must conclude that every man, woman, and child in the United States had 1/60th of a baby last year." Needless to say, the denominator of the ratio could have been selected more judiciously. Here are some more examples:

You can multiply the number of felons in penitentiaries by 365, and then average this out among the some 210,000,000 million people in the United States and discover that the "average person" serves about 10 hours a year for a felony.

". . . last year 675 million marijuana cigarettes were smoked in the U.S. —$3\frac{3}{8}$ for each man, woman, and child—in a flaunting of the law rivaling Prohibition."[8]

[7] Earle S. Newman, "Intermodal Freight Transportation in the United States," *Public Roads*, December 1968, p. 110.

[8] *Popular Science Monthly*, May 1968, p. 76.

"Pulling the drawstring tighter, the men's pajama industry dolefully reported some raw facts last week: Men were buying only one-third of a pair of pajamas each year."[9]

The Black-or-White Fallacy

When logicians speak of the black-or-white fallacy they refer to the error committed when it is supposed that for a controversial question, there are two alternative, and diametrically opposite, solutions and no more. We are particularly vulnerable to this fallacy when we think in terms of extremes. For in addition to the extremes, there are usually several intermediate positions or alternative courses of action. Sometimes statistical information is presented in such a way that we are encouraged to think only in black-or-white terms. For example, a government spokesman back in the days of the Johnson Administration proclaimed that the San Francisco referendum on the Vietnam War showed overwhelming support for the Administration's policies. Thirty-seven percent of the electorate voted for and 63 percent voted against a position which called for immediate withdrawl of American troops from Vietnam.[10]

The alternatives offered in this case were extremes—get out of Vietnam immediately or support the Administration's existing policies. But just because one doesn't favor getting out immediately, it doesn't necessarily follow that he supports existing policies. He may prefer a third, less extreme, course of action.

Equivocation

Many a comedian has extracted a laugh from his audience through an old joke about freight carloadings. The comedian holds up a large chart much like the one shown in Figure 14-1:

"The broken line," he explains, "represents the number of freight carloadings in the United States during the past seven years. The solid line represents the number of gallons of alcoholic beverages consumed in the United States during the same period. These figures," he concludes with a professorial air, "positively prove that more people are getting loaded than freight cars."

The fancy word used to describe the logical fallacy intentionally committed by the comedian is *equivocation*, which means that a specific word is used in one sense at a given point in an argument and in quite a different sense

[9] *Newsweek*, August 1, 1949, pp. 50–51.
[10] *The New York Times*, November 26, 1967.

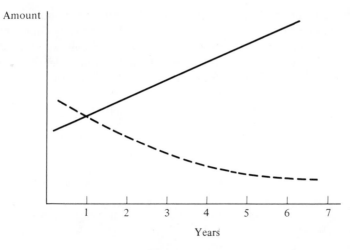

Figure 14–1

at a later point. Clearly, a loaded person is loaded in quite a different way than is a loaded freight car.

Although this fallacy isn't committed terribly often in conjunction with statistical data, it does crop up now and then, usually quite unintentionally. For example, an antilabor man might say, "National statistics tell us that four percent of the labor force is *unemployed*, but I can attest to the fact that the true percentage is much higher than that. Because of union-encouraged featherbedding, a large proportion of workers drawing big pay are *unemployed* a large part of the time." The national statistics referred to represent the proportion of people in the labor force not on company payrolls but wishing they were. To be unemployed in this sense, therefore, would entail not being on a payroll of some kind. To be unemployed in the second sense would entail standing around doing little or nothing for one's pay—that is, to not be employed in productive activity.

A speaker might assert, "We are told that prostitution is a growing national problem, but that isn't the half of it! At least half of the men and women in this country today are prostitutes. They sell their bodies or their minds on jobs that are personally meaningless and socially destructive." Obviously, the word "prostitute" is being used in two different senses. The first time it is used, it is meant literally; the second time, only figuratively.

Undoubtedly a number of relatively common statistical fallacies have been overlooked. Still, we have now dealt with quite a long list of these hindrances to straight thinking. If you can recognize the ones we *have* covered when you encounter them in books, speeches, and newspaper and magazine articles, you can justifiably consider yourself a pretty sophisticated critic of statistical evidence. If you are the brave sort and wish to test your new-found acumen, you might try analyzing some or all of the carefully selected examples of statistical arguments found in the next chapter. If you prefer to assume that because you have now read a book on statistical fallacies you are adequately prepared to cope with statistical evidence as it crops up in the real world, then I thank you for your attention and wish you good luck.

15

Assorted Examples to Think About

There is no merit where there is no trial; and till experience stamps the mark of strength, cowards may pass for heroes, and faith for falsehood.

—A. HILL

The time has come for you to try your own wings. Many of the following examples are clearly fallacious—some in more than one way. Many others are not necessarily fallacious but do include material which should give the statistical critic pause. All are well worth analyzing. Let's see what you can do with them.

A marketing research study uncovers this valuable fact: The average age at which young men in Fuzzville, U.S.A., begin shaving is 16.537 years.

> Many VA employes justify the inefficiency, the prolonged hospitalizations and the unnecessary admissions by saying that it costs but a few dollars a day to hospitalize a patient in the VA. This is not true. Our VA-hospital finance office calculates that to hospitalize one patient costs the taxpayer $28.02 per day—without including the costs of hospital building and equipment depreciation. Even using this low figure one may calculate that the taxpayers pay $933.07 for the average VA hospitalization; in a private institution, at a higher daily rate but with a much shorter average stay, the patient would have paid about $400.[1]

[1] Richard S. Dillon, "Bureaucratic Medicine: A View of Veterans' Hospitals," *Atlantic Monthly*, November 1962, p. 81. © by the Atlantic Monthly Company, Boston, Mass. Reprinted with permission.

A sorority dance was considered a huge success by its planners because only four people out of the 200 in attendance complained. "When only four people are discontent and 196 are delighted," one of the planners was heard to say, "as far as I'm concerned, that's a successful dance."

According to a survey over 75 percent of the people questioned know couples who engage in wife swapping. The conclusion was reached that about 75 percent of the married couples in the United States engage in wife swapping.

> . . . So while TV is important, it won't do everything, or reach everybody. For instance, with television you can't be sure you're getting through to the upper-income, college-educated group. That's the group that spends more on groceries every week. 81% of them aren't avid watchers. But they are eager readers. And the higher the education and income, the more eager magazine readers you find.[2]

> In time-utilization studies it is the rule and not the exception to find most people working 20% to 40% below capacity. And I am defining capacity as the efficiency they would comfortably manage without additional effort and time.[3]

> . . . The figures speak for themselves: of the lower class couples interviewed by _____, only 44 per cent of the husbands and 20 per cent of the wives expressed "great interest and enjoyment" in intercourse. But with the middle classes, the percentages shot up sharply to 78 per cent for the husbands and 50 per cent for the wives.[4]

The above facts were obtained from a study of 152 married couples in the midwestern United States.

A newspaper article asserted that children do not copy their parents' smoking habits. According to a survey including some 50,000 children aged 11 to 18, parents' smoking habits had almost nothing to do with whether the children smoked. Among smokers, 70 percent had fathers who smoked; but 63 percent of the nonsmokers also had smoking fathers.

A magazine article asserted that 75 to 80 percent of all convicted criminals are products of broken homes or unhappy childhoods.

> The George Washington University survey classified 56% of all drinkers as moderate, only 12% as immoderate.

[2] Magazine Publishers Association. (Advertisement)

[3] "Practical Solutions to Management Problems," *Business Management*, April 1968, p. 58.

[4] *Newsweek*, April 17, 1967, p. 100.

All of the evidence, in fact, sustains the conviction that the average American knows how to handle his liquor.[5]

. . . a particular product of our civilization and of our carelessness is simply human litter and it is heaved into the atmosphere by all of us—from automobile tail pipes, from incinerator stacks, from power stations and home chimneys. Last year over 143 million tons of it went up and fell right back down on beautiful America . . .[6]

Alcoholism costs U.S. businesses $4.3-billion annually, according to a recent study by the National Council on Alcoholism, Inc. Others, citing botched deals and lost sales at high executive levels, feel this estimate is too low and peg the annual cost closer to $7-billion.[7]

An executive of a men's apparel manufacturing company was asked by a superior to compute an estimate of the number of men's suits sold in the United States in a year's time. He proceeded by asking all male employees of this specific company how many suits they typically buy in a year. The resulting total was multiplied by official Census Bureau estimates of the number of companies in the United States. He then phoned ten of his acquaintances in the professions—teachers, doctors, lawyers, etc.—and asked them the same question. This total was multiplied by one-tenth of the official estimate for number of professionals in the country. The two results were then summed. This sum was the estimate our executive submitted to his superior.

A certain company has 1,900,000 shares of stock outstanding and 3,700 common stock holders two of whom own 1,000,000 shares between them. In the annual report the statement is made: "Our common stock holders own an average of almost 514 shares apiece."

A certain football team has six members who weigh 240 pounds, eight who weigh 225 pounds, three who weigh 200 pounds, and three who weigh 180 pounds. Their average weight is reported by the coach as 211.25 pounds.

A list of best-selling books is made up by investigating several stores reporting. The best seller is then stated to be the book which most stores report selling most of.

An employee of a certain company was about to be transferred to another state and had his choice between Kentucky and California. Since climate was a matter of considerable importance to him he did some investigating into the temperatures of the two states. He learned that, over the course of a year, the

[5] *Time*, December 29, 1967, p. 15.

[6] *Life*, February 7, 1969, pp. 38–39.

[7] *Business Week*, October 26, 1968, p. 97.

two states have about the same average temperature. He concluded, therefore, that insofar as temperature was a factor, it didn't make any difference to which state he was transferred.

It is imperative that a certain kind of plastic tubing be no less than 12 inches in length. A quality control engineer examines a particular batch and finds that the average length is 15 inches. He concludes that the batch is satisfactory.

> The advertisers who buy Godfrey seem to stay awhile. His 1970 sponsor list of 27 companies represents an aggregate of 86 years of participation, an average of more than three years for each company.[8]

Tom, Dick, and Harry got appreciably higher scores on the final examination for a certain course than they achieved on the midterm examination. Tom's score increased by 25 percent, Dick's by 20 percent, and Harry's by 15 percent. Tom asserted that, among them, they had raised their scores by 60 percent (25 + 20 + 15 equals 60 percent). Dick argued that the correct percent was 18.3 (25 + 20 + 15 divided by 3 equals 18.3 percent). Harry maintained that Tom and Dick were both wrong.

A man's income increased from $4000 during his last year in night school to $8000 the following year—an increase, he figured, of 50 percent.

A certain company experienced a 23 percent increase in its costs of production and only a 15 percent increase in selling price during a certain five-year period. The company's chief negotiator lamented to the head of the employees' union that this 8 percent decrease in profits was putting the very future of the company in jeopardy. He urged the union, on the basis of the preceding argument, to lower its wage demands.

A manufacturer registered the following complaint with a government agency:

> We have been hard hit by material allocations. As a matter of cold fact there has been a 150 per cent decline in our production in this quarter compared to the same quarter last year. You know what this does to costs.[9]

According to a magazine article, the lung-cancer mortality rate for women has risen by 400 percent in the past 40 years—and 1700 percent for men.

Cargill thinks that the laws against use of marijuana should be made more severe. He asserts, "95 percent of heroin addicts started on marijuana."

[8] CBS Radio Network

[9] Cited in William Addison Neiswanger, *Elementary Statistical Methods*, Revised Edition (New York: The Macmillan Company, 1956), pp. 45–46.

Munster, on the other hand, thinks the laws are already too strict and points out that fewer than 3 percent of marijuana users go on to use the harder drugs.

During a Senate debate on extending the draft, one Senator reported that members of the U.S. Congress have some 230 draft-eligible sons, but over one-half were "waivered out for one reason or another; and only 26 have been sent to Vietnam. The Senator concluded that these figures prove that the Selective Service favors upper-income groups.

There is increasing evidence, however, that the potentially damaging motivations—money and prestige—are beginning to outweigh humanitarianism as the professional *raison d'être*. A study by the Student American Medical Association, reported in GP magazine, surveyed 694 medical students, interns, residents and prospective students, with dismaying results. In response to the survey, which in effect asked: "What were the major reasons you chose medicine as a career?," 56 percent answered "income" or "prestige" rather than "humanitarianism." Dr. _____ is one of many who are frankly concerned. "Their ambition is not achievement but financial security," he laments. "It bodes no good for the standards of our profession."[10]

Women outdrive men which is shown in the following statistics:

(1) Of men involved in accidents 23% had been drinking—against 9.6% for the women.

(2) One third of the men were driving too fast, one fourth of the women.

(3) Of 101 drivers involved in an accident while passing on a curve, 15 were women.

Going further, the patrol looked into bicycle accidents. Some 3000 males were injured on bicycles in the state in 1967 and 34 were killed compared with 662 females injured and 11 killed.[11]

An advertisement for a driving school boasts that whereas The Harvard Law School graduates 1 out of 6 applicants, this particular driving school graduates only 1 out of 10 applicants.

A magazine article about the Washington Senators professional baseball club indicated that the club was moving its franchise from Washington to Dallas-Fort Worth because they were losing money in Washington. To emphasize this fact, the article quoted the home game attendance figures:

As the Senators disintegrated from a mediocre but interesting ball club to a bad but uninteresting one (they are currently 33 games out of first place in

[10] Martin L. Gross, *The Doctors* (New York: Dell Publishing Co., Inc., 1966), p. 28. © Random House, Inc., Alfred A. Knopf, Inc.

[11] *The Rocky Mountain News* (Denver, Colorado), November 27, 1968.

the league's Eastern Division), home attendance dropped from 824,789 last year to 631,933 so far this season.[12]

It was found that only 14% of the divorce cases in this country involve men who had married between the ages of 25 and 30 and that only 11% involve men married between the ages of 30 and 35. These figures were taken as proof that the older a man is when he marries, the more likely the marriage is to be successful.

"Graduates can afford to be choosy about careers these days. For one thing, there are so many more occupations—21,741 at the last count by the Department of Labor."[13]

There are 48 million Americans who call themselves Roman Catholics. . . . From a nationwide survey of U.S. Catholics above the age of 17, . . . , a strikingly bleak picture emerges. More than a third do not attend Mass regularly.[14]

The above information was interpreted as an indication that many Catholics are falling away from their religion.

In the U.S., which takes 50% of the global coffee output, per capita daily consumption has fallen from 3.12 cups to 2.86 cups in four years, apparently because younger Americans tend to prefer soft drinks.[15]

It was determined from a reputable study that 75 percent of the people using Sparkle-Plus Toothpaste for a trial period developed no cavities. This finding was written up in a newspaper article entitled "Nonusers of Sparkle-Plus Get Cavities!"

Bernie Skiles hit two home runs his first two times at bat during a certain game. As he approached the plate for the third time, a spectator was heard to say, "Old Bernie won't get a home run this time; the odds against three home runs in a row are astronomical."

A recent divorcee comforted himself by saying, "My next marriage is sure to work. The odds against two divorces are huge."

In a certain city, the average number of school-age children per family having school-age children was estimated by questioning a sample of children in schools. The figure obtained was much too high. Why?

[12] *Time*, October 4, 1971, p. 60.

[13] *Time*, May 30, 1969, p. 42.

[14] *Newsweek*, October 4, 1971, pp. 80–81.

[15] *Time*, September 1967, p. 61.

1,189 psychiatrists say Goldwater is psychologically unfit to be president! This statement appeared on the cover of *FACT* magazine one month before the presidential election of 1964. This data was the result of a questionnaire sent to the 12,356 psychiatrists listed by the American Medical Association. Of the 2,417 who replied, 657 said Barry Goldwater was fit for the presidency, 571 declined to take a position, and 1,189 called him unfit.[16]

A study found that although the severe poverty group and the comfortable groups differ in income, family structure, etc., they are quite similar in the value they attach to educational goals.
The study was conducted by questioning a random sample of all students in 11 colleges in the Deep South.

Marketing researchers for a certain automobile company prepared a questionnaire for Panther owners (a small sports car manufactured by this company) to determine the characteristics these owners prefer in future Panther models. The questionnaire was completed by all participants of a Panther Club national convention. The results of this questionnaire were taken into account when future models were designed.

My answer is this: Although proving cause and effect relationships in sociological matters is difficult, common sense tells us that it is ridiculous to imply that pornography has *no* effect. Since the current flood of erotica began in the early 1960's, sex crimes have multiplied. From 1960 through 1969, reported rapes increased 116 percent; arrests for rape went up 56.6 percent; and arrests for prostitution and commercialized vice shot up 60 percent. Such statistics at least appear to reflect some "significant" relationship between crime and pornography. . . .[17]

Of 51 battered-infant cases studied over a nine-year period, Drs. _____ and _____ of the Montreal Children's Hospital in Canada found that 12 of the infants, or 23.5%, were infants who weighed below normal at birth.
Since only 7 to 8 percent of infants born in Quebec are low in weight, the two physicians concluded in the July *American Journal of Diseases of Children* that low birth weight may well contribute to child-beating.[18]

An article entitled "They Put a Parson on the Payroll" in a popular magazine states: "In just two years religion-on-the-job has accomplished several pretty wonderful things . . . labor turnover has dropped from 7.61 to 5.22% in two years, the accident rate has declined approximately 40%, and absenteeism is much lower than it used to be."[19]

[16] *Time*, May 17, 1968, p. 52.

[17] Charles H. Keating, Jr., "The Report That Shocked the Nation," *Reader's Digest*, January 1971, p. 37.

[18] *National Enquirer*, October 24, 1971, p. 29.

[19] Cited in William A. Spurr and Charles P. Bonini, *Statistical Analysis for Business Decisions* (Homewood, Illinois: Richard D. Irwin, Inc., 1967), p. 12.

"National Retail Merchants Association says the Pill is responsible for a 19% decline in sales of maternity clothes last year."[20]

U.S. STATISTICS PROVE IT . . . MARRY AND LIVE LONGER:
There's no question about it: Married people live longer than single people; and people who were once married live longer than people who were never married.

The Public Health Service's National Office of Vital Statistics has proved this to be a fact. Basing his figures on the 1950 census and mortality rates for 1949, 1950, and 1951, statistician _____ showed that deaths among bachelors were almost two-thirds greater than among husbands.

Among divorced men and widowers, the rate was half again more.[21]

The workers behead the huge "Fascist" puppet and plan a democratic Italy. But the new tyranny becomes the assembly line, about which Fo raises a characteristically Italian plaint: "Women who work on the assembly line are forced to make 40,000 body movements a day. As a result, 15% of them become sterile and 30% cripples. In some factories where men are subjected to continual movement and noise, 40% of the men become impotent."[22]

A study of medical records of 352,000 people was made and results were analyzed to determine factors which may contribute to death by stroke and heart attack. Death rates were 40–53% higher for men and women under 50 who had slept over 7 hours a night regularly. For women over 70, the rate was 167% higher, and it was 286% higher for men between 50 and 59.

This article implied that the best amount of sleep was 7 hours a night or less.[23]

A Columbia University's Neurological Institute in the last year before the pill came into wide use, only two women between the ages of 20 and 40 were admitted with strokes; one was pregnant, and the other case was apparently unrelated to hormone changes. Last year, Dr. _____ and Dr. _____ report, they had nine such cases in this age range, and all but one of these patients were on the pill. Some had been taking them for years, some only for a few weeks.[24]

A survey conducted at a certain college revealed that only a minority of girls were in favor of remaining virgins until the day they marry. The college newspaper wrote up the results under the headline "Most Coeds Favor Promiscuity."

A teacher noted that many students who had low scores on the midterm examination were much closer to class average on the final examination. This she attributed to her teaching skill. But she also noticed that several

[20] *Newsweek*, November 25, 1960, p. 83.

[21] *Chicago Daily News*, April 8, 1955.

[22] *Time*, March 21, 1969, p. 72.

[23] *Time*, October 25, 1968, p. 64.

[24] *Time*, December 29, 1967, p. 33.

students who were exceptionally high on the midterm examination slumped noticeably on the final. This she attributed to slackening off due to over-confidence.

An advertisement began "One out of three people you work alongside is a law breaker of some sort!"

Most doctors know that visitors often do more to stir up hospital patients than to soothe them. But the doctor's own ward rounds can have the same effect, sometimes with fatal results, reported Finnish Doctor _____ in the *British Medical Journal*.

Studying the histories of 39 Helsinki hospital patients who died of coronary occlusion after stays of seven to 42 days, Dr. _____ discovered that six of them, subject to severe emotional stress, had died during or after a phy-sicians's visit. Among the cases:

An accountant, 58, came to the hospital 21 days after an an attack of angina pectoris. He seemed in satisfactory condition until the 16th day in the hospital. The head physician was making his round; as the doctor drew closer, the patient became nauseated, suffered a severe attack and died within two hours.

After suffering chest pains during a tantrum, a female post-office clerk, 68, was admitted for treatment. In the ward, she grew excited over trivialities. After nine days, when the doctor approached she became restless. Asked how she felt, she tried to answer, and died on the spot. . . .[25]

THE END

[25] *Time*, February 21, 1955, p. 37.

Index